ENVIRONMENTAL SCIENCE, ENGINEERING AND TECHNOLOGY

INDOOR AIR AMBIENCES

Environmental Science, Engineering and Technology

Additional books in this series can be found on Nova's website under the Series tab.

Additional E-books in this series can be found on Nova's website under the E-books tab.

Public Health in the 21st Century

Additional books in this series can be found on Nova's website under the Series tab.

Additional E-books in this series can be found on Nova's website under the E-books tab.

ENVIRONMENTAL SCIENCE, ENGINEERING AND TECHNOLOGY

INDOOR AIR AMBIENCES

JOSÉ A. OROSA GARCÍA

Nova Science Publishers, Inc.
New York

Copyright © 2011 by Nova Science Publishers, Inc.

All rights reserved. No part of this book may be reproduced, stored in a retrieval system or transmitted in any form or by any means: electronic, electrostatic, magnetic, tape, mechanical photocopying, recording or otherwise without the written permission of the Publisher.

For permission to use material from this book please contact us:
Telephone 631-231-7269; Fax 631-231-8175
Web Site: http://www.novapublishers.com

NOTICE TO THE READER

The Publisher has taken reasonable care in the preparation of this book, but makes no expressed or implied warranty of any kind and assumes no responsibility for any errors or omissions. No liability is assumed for incidental or consequential damages in connection with or arising out of information contained in this book. The Publisher shall not be liable for any special, consequential, or exemplary damages resulting, in whole or in part, from the readers' use of, or reliance upon, this material. Any parts of this book based on government reports are so indicated and copyright is claimed for those parts to the extent applicable to compilations of such works.

Independent verification should be sought for any data, advice or recommendations contained in this book. In addition, no responsibility is assumed by the publisher for any injury and/or damage to persons or property arising from any methods, products, instructions, ideas or otherwise contained in this publication.

This publication is designed to provide accurate and authoritative information with regard to the subject matter covered herein. It is sold with the clear understanding that the Publisher is not engaged in rendering legal or any other professional services. If legal or any other expert assistance is required, the services of a competent person should be sought. FROM A DECLARATION OF PARTICIPANTS JOINTLY ADOPTED BY A COMMITTEE OF THE AMERICAN BAR ASSOCIATION AND A COMMITTEE OF PUBLISHERS.

Additional color graphics may be available in the e-book version of this book.

LIBRARY OF CONGRESS CATALOGING-IN-PUBLICATION DATA

Orosa Garema, Josi A.
 Indoor air ambiences / Josi A. Orosa Garema.
    p. cm.
 Includes bibliographical references and index.
 ISBN 978-1-61209-570-7 (softcover)
 1. Buildings--Environmental engineering. I. Title.
 TH6021.O76 2011
 613'.5--dc22
                    2010051731

Published by Nova Science Publishers, Inc. ┼ New York

DEDICATED TO MY PARENTS, BROTHERS AND SISTER

Contents

Preface		ix
Abstract		xi
Introduction		xiii
Section 1:	**Indoor ambiences background**	1
Chapter 1	Background	3
Chapter 2	Introduction to Thermal Comfort	15
Chapter 3	Introduction to Indoor Air Quality	45
Section 2:	Research techniques	65
Chapter 4	Materials and Methods	67
Chapter 5	Software Resources	81
Section 3:	**Practical Case Studies**	99
Chapter 6	Indoor Air and Health Effects: A Practical Case Study of Flats	101
Chapter 7	Indoor Air and Energy Saving: A Practical Case Study of Offices and Schools Buildings	113
Chapter 8	Indoor Air and Materials Conservancy: A Practical Case Study of Libraries	131

Chapter 9	Indoor Air And Work Risk: A Practical Case Study of Industrial Environment	**147**
Thankfulness		**157**
Index		**159**

PREFACE

This new book presents an overview of conclusions and future works in new research and design parameters of Spanish indoor ambiences. These conclusions are based on real data conducted at the Department of Energy at the University of A Coruña. This research involved a set of ambiences such as flats, schools, libraries and ships located in the area of A Coruña, Spain. The purpose of these investigations was to determine the correlation between indoor ambiences, comfort, energy saving, work risk, health and material conservancy.

ABSTRACT

This book represents an overview of conclusions and future works in new research and design parameters of Spanish indoor ambiences. These conclusions are based on real data measured in different research conducted at the Department of Energy at the University of A Coruña. This research involved a set of ambiences such as flats, schools, libraries and ships located in the area of A Coruña, Spain. The purpose of these investigations was to determine the correlation between indoor ambiences, comfort, energy saving, work risk, health and material conservancy.

The general conclusion obtained regarding air exchange in flats was that, given the current levels of occupation, ventilation procedures should be modified to keep the relative humidity lower than the maximum recommended level of 60%. In this case, natural ventilation could be effectively used in keeping both indoor temperature and humidity, as well as the microbiological load within acceptable limits. On the other hand, in libraries with natural ventilation, material conservancy can improve with mechanical ventilation.

Finally, old schools with higher thermal inertia showed better indoor conditions than new ones and industrial ambiences in ships showed extreme temperature related with higher work risk that could be reduced with an increase in air change rate.

INTRODUCTION

This book is a compilation of the author's research activities with the intended proposal to show the main findings and techniques learnt over the last years of research activity in these areas and facilitate the learning of future researchers in this field. Such activity has been directed to study indoor environment conditions and their effect on the conservation of materials (museums), energy optimization (banks), risk (vessels), diseases (hospitals) and improving comfort conditions (housing).

All these aspects have enabled the development of related doctoral dissertations since 1998 and participation in multiple international congresses and meetings led by the International Energy Agency. This international activity poses a special interest for Spain since this group is the only Spanish representative at the IEA meetings.

The book is structured into several chapters. The first to the third chapters describe the research activity in areas of air-conditioning and air treatment that have been developed worldwide up to now. The fourth and fifth chapters show the main measuring apparatus, theoretical models and software needed to carry out such investigations, while the remaining chapters show the main procedures and conclusions developed in four different environments: homes, schools, museums and ships.

The author;
Dr. José Antonio Orosa García

SECTION 1: INDOOR AMBIENCES BACKGROUND

INTRODUCTION

In this first section it will be showed the principal research activities in the area of thermal comfort and indoor air quality in the last years. This activities will let to reader have an idea of actual methods and its origins. In consequence, this historical revision is considered necessary to an adequate comprehension of objectives of the most actual research projects and its methodologies employed.

Chapter 1

BACKGROUND

INTRODUCTION

Given the varied activities of international involvement in indoor environments it is necessary to start this book with a chronological development of the main researches and conclusions reached until now in general and local thermal comfort and in the perception of the indoor air quality area.

1.1. GENERAL THERMAL COMFORT BACKGROUND

Fanger's PMV model is based on thermoregulation and heat balance theories. According to these theories, the human body employs physiological processes in order to maintain a balance between the heat produced by metabolism and the heat lost from the body.

In 1967 Fanger investigated the body's physiological processes when it is close to neutral to define the actual comfort equation. His investigations [1] began with the determination that the only physiological processes influencing heat balance were sweat rate and mean skin temperature as a function of activity level. After this, he used data from a study by McNall et al. (1967) [2] to derive a linear relationship between activity levels and sweat rate and conducted a study to derive a linear relationship between activity level and mean skin temperature. These two linear relationships were substituted into heat balance equations to create a comfort equation to describe all combinations of the six PMV input variables that result in a neutral thermal sensation.

Having obtained an initial comfort equation, it was validated against studies by Nevins et al. (1966) [3] and McNall et al (1967) [2], in which college-age participants rated their thermal sensation in response to specified thermal environments. To consider situations where subjects do not feel neutral, the comfort equation was corrected by combining data from Nevins et al (1966), McNall et al (1967) and his own studies, Fanger (1970) [4]. The resulting equation described thermal comfort as the imbalance between the actual heat flow from the body in a given thermal environment and the heat flow required for optimum comfort (i.e. neutral) for a given activity. This expanded equation related thermal conditions to the seven-point ASHRAE thermal sensation scale, and became known as the PMV index. Fanger (1970) also developed a related index, called the Predicted Percentage Dissatisfied (PPD). This index is calculated from PMV, and predicts the percentage of people who are likely to be dissatisfied with a given thermal environment.

Thermal comfort standards use the PMV model to recommend acceptable thermal comfort conditions. The recommendations made by ASHRAE Standard 55 [5] are shown in Table 1.1.1. These thermal conditions should ensure that at least 90% of occupants feel thermally satisfied.

Table 1.1.1. ASHRAE Standard recommendations

	Operative Temperature	Acceptable range
Winter	22°C	20-23°C
Summer	24.5°C	23-26°C

These conditions were assumed for a relative humidity of 50%, a mean relative velocity lower than 0.15 m/s, a mean radiant temperature equal to air temperature and a metabolic rate of 1.2 met. Clothing insulation was defined as 0.9 clo in winter and 0.5 clo in summer. All these concepts such as operative temperature and clo will be explained in chapter 2.

1.2. LOCAL THERMAL COMFORT BACKGROUND

It was in 1956 that the first serious studies on local thermal comfort background began, when Kerka and Humphreys began their studies of indoor environment. However, ever since, man has had a special interest in controlling indoor environments.

In these studies they init to use panels to assess the intensity of smell of three different fumes and smoke to snuff. The main findings show that the intensity of the odour goes down slightly with some increase in atmospheric humidity. Another finding obtained, indicates that in the presence of smoke snuff, the intensity of the odour goes down with increasing temperature, for a constant partial vapour pressure.

In 1972, Fanger [4] described the general thermal comfort as a material and energy balance being the basis of the actual general thermal comfort research. The ISO 7730 and the ASHRAE Standard 55-2004 reflected this theory. Two years later, Cain [7] explored the adaptation of man to four air components and to different concentrations over a period of time. The main conclusions obtained showed that there was no significant difference between pollutants. In all of these, perceptions often fell by 2.5%/s initial value until reduced to 40%.

In 1979, Woods [8] confirmed the results of Kerka and concluded that smell perception of odour intensity is linearly correlated with the enthalpy of air.

In 1983, Cain et al.[9] studied the impact of temperature and humidity on the perception of air quality. They concluded that the combination of high temperatures (above 25.5 °C) and a relative humidity above 70% exacerbate odour problems. Six years later, Berglund and Cain [10] discussed the adaptation of pollutants over time for different humidities. This study concluded that the air acceptability, for different ranges of humidity at 24 °C, is stable during the first hour. It was also concluded that the subjective assessment of air quality was mainly influenced by temperature conditions and relative humidity and, secondly, by the polluted air. It was also concluded that the linear effect of acceptance is more influenced by temperature than by relative humidity.

In 1992, Gunnarsen [11] et al. studied the possibility of adapting the perception of odour intensity. Such adaptation was confirmed after a certain time interval.

In 1996, Knudsen et al. [12] carried out research into the air before accepting a full body and facial exposure. The problem with this test is that the process is carried out at constant temperature equal to 22 degrees Celsius and the relative humidity is not controlled.

In 1998 L. Fang, Clausen and P.O. Fanger [13] carried out an initial experiment in a chamber with clean air heated to 18 °C and 30% relative humidity, see Figure 1.2.1. In the experiment, 40 subjects without specific training were subjected to the conditions in these chambers. As a precaution, they were warned not to use strong perfumes before the experiment. The subjects underwent a facial exposure; they were questioned about their first impression of the air quality inside the chamber. In this case we consider the existence of clean air where there are no significant sources of pollution and the air has not been renewed with outdoor air.

From these studies it was concluded that there is a linear relationship between the acceptability and enthalpy of the air. It also concluded that, at high temperature levels and humidity, the perception of air quality appears more influenced by these variables than by the air pollutants. All these findings need further validation, which involved the development of more experiences.

In a second experiment this same group carried out a study of the initial acceptability and subsequent developments. For this study they used clean air and whole body exposure of individuals to different levels of temperature and humidity. This experiment was divided into two sets: one aimed at defining the feeling of comfort and the other at defining the perception of smell.

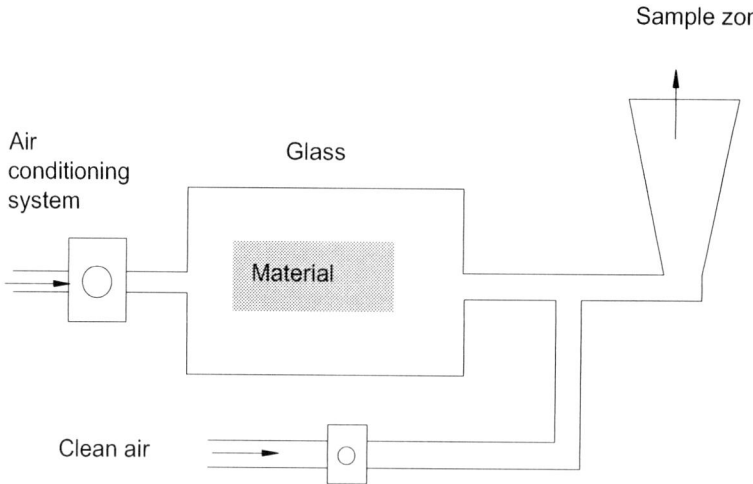

Figure 1.2.1. House heated Climpaq designed by Albrectsen in 1988.

For these experiments a system was developed based on two stainless steel chambers (3.60 x 2.50 x 2.55 m) independent and united by a door that allows a camera to pass from one to the other. In this way, the individual who performs the test may turn to the second chamber at each stage of the experiment. This camera is subjected to a new odour level, temperature and/or humidity. Figure 1.2.2 shows the shape of the chamber.

The experiment focused on conducting a survey on 36 students who had not been trained in issues of indoor environments.

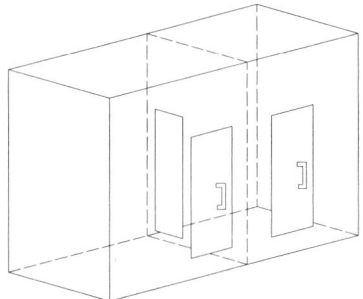

Figure 1.2.2. New experimental chamber.

The group was made up of 26 men and 10 women. All of them were nearly 25 years old and had their whole body exposed in the chamber.

The scale of values that were employed during the survey is shown in Figure 1.2.3.

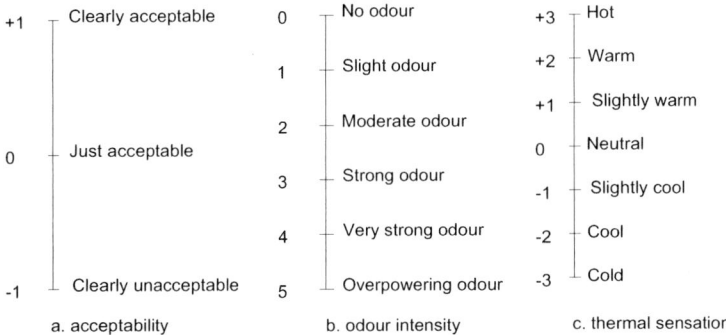

Figure 1.2.3. Used survey.

In these chambers there remained constant different temperatures and humidity within the ranges (18-28 °C) and (30-70%), respectively. The number of air changes in both chambers was the same and equal to 420 l/s.

The existing pollutants came from the own chamber or from the air renovation system.

Every 20 minutes existing conditions were varied which prompted the individual to change camera. The questionnaires were filled in every 2.5, 5, 10, 15 and 20 minutes. Throughout the process the subjects could adapt their clothing to the environment around them to achieve thermal neutrality.

In the second round of experiments individuals were submitted to a similar procedure to the previous one. In this case a contaminated source like PVC was introduced, and air renovation descended to 200 l/s. The pollutants were hidden in the camera and individuals were introduced in groups of six to answer the survey. The findings for the first experiments indicated that the alarm immediately jumps depends on the temperature and relative humidity in the new chamber. It is also concluded that the alarm after 20 minutes does not depend on the conditions of initials temperature and relative humidity.

The results show that there is an increasing acceptability with the drop in temperature and relative humidity and that a cooling of the mucous membranes is essential to perceive the air as acceptable, as it demonstrates the influence of the air enthalpy. The results indicated that, for a whole body exposure, there is a linear relationship of the acceptability with the enthalpy, both for clean air as polluted, see Figure 1.2.4. It was concluded that there is no difference between the initial acceptability and the acceptability after 20 minutes of exposure. It also follows that the acceptability is independent of the environment conditions that surrounds the individual before entering the camera.

The findings of tests on odours indicate that the intensity of the odour varies little with temperature and relative humidity and that there is some adjustment to smell after about twenty minutes. Berglund and Cain studies (1989) were proved in the absence of adaptation of acceptability in time. It also checks the conclusion of Gunnarsen (1990) when it confirmed adaptation to smell inside after some time.

An important conclusion that has been reached with this review is that it is possible to save energy if you lower the number of air changes, temperature and relative humidity. These discussions are ongoing to maintain the PD with the corresponding energy savings. We must remember that cold, very dry air with high pollution causes the same number of dissatisfaction than clean, mild and more humid air. It is interesting to note that if the temperature and elative humidity slightly drop, pollutants emitted by each of the materials (Fang 1996) will be reduced so.

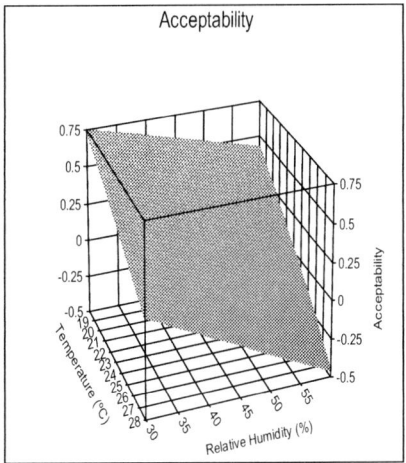

Figure 1.2.4. Influence of temperature and relative humidity on the acceptability.

At present, field tests are recommended by the majority of researchers so they can perform characterization of environments according to their varying temperature and relative humidity. This may start the validation of models that simulate these processes by computer.

1.3. INDOOR AIR QUALITY BACKGROUND

Airborne micro organisms may be responsible for infectious or allergic disorders such as asthma and humidifier fever in exposed people [14]. Numerous epidemiologic studies in the past two decades provide a dose-response correlation between dust mite exposure, with mite sensitisation, and to some degree, asthma [15]. There seems to be insufficient evidence for the associations of indoor mould with asthma [16].

Otherwise, at present there is no conclusive evidence for the role of indoor fungi for allergies in children. Some studies find association, others do not. More than 80 genera of fungi have been associated with symptoms of respiratory tract allergies (Horner et al. 1995) [17, 18]. Results of [16] lead to the conclusion that with current methods for assessing mould exposure and allergy history it remains difficult or impossible to determine causality and attributable health risks.

There is also a need for improvement and standardization in the tools for exposure, assessment of moulds at genus level, fungal antigens, fungal fragments and metabolites. In consequence, prospective longitudinal and intervention studies are required to further elucidate the role of fungi in environmental hygiene and allergies [16].

In this sense, little data is currently available to guide the clinician on primary prevention of allergies. However, recommendations such as encouraging breast-feeding and allowing the child to grow up in a smoke-free environment remain valid. Allergen avoidance is crucial for treating and preventing disease exacerbation in allergic and already sensitised patients. In addition to recommending allergen avoidance as a therapeutic approach, allergologists have issued recommendations to rigorously avoid contact with potential allergens, such as certain foods during infancy and pets, also for primary prevention of allergic disease, i.e. for prevention of sensitisation [18].

If indoor allergen exposure is a principal determinant of asthma, then at the extremes of allergen exposure, we might expect to find: (1) lower asthma prevalence in locales without mite or pet allergen; and conversely, (2) higher asthma prevalence in locales with high indoor allergen exposure [15].

However, things are more complicated and are not the same for all allergens: while increased exposure to house dust mite allergen is paralleled by increased sensitisation rates, the same is not true for cat allergen. One possible explanation for the different effects of different allergens may be their biochemical properties: mite allergens, in contrast to cat and dog allergens, contain proteolytic enzymes. It has been shown that in regards to house dust mites a low-allergen environment can be achieved [19]. Other factors such as the distance of a building from the source (a nearby park) and supermicrometre particle concentrations will be associated with the concentration levels of fungi [18].

At the present time, the only way to guarantee lower mite allergen levels in modern homes in the western world is to remove carpets and to encase the mattresses and bedding [20]. Furthermore, to reduce this exposure we must improve IAQ. The 3 primary considerations in improving IAQ are (1) evaluation of construction failures that allow moisture into the walls of a building, (2) poor ventilation causing excessive humidity and accumulation of gaseous and/or chemical exposure from materials in the living space, and (3) poorly designed or failing HVAC systems that contribute to poor air calculation.

In regards to these two last points, some authors [14] have concluded that massive proliferation of microorganisms may take place in an HVAC unit with certain risk factors such as low efficiency filters, cold mist humidifiers using water recycling, areas in which condensation water remains stagnant, large recirculation of air and faulty or deficient maintenance conditions. They demonstrated that, compared to a naturally ventilated building, a HVAC system which is well-designed and well-maintained improves the microbiological quality of indoor air.

Finally, not all indoor allergens are necessarily equal in their propensity to cause asthma. For example, using dust allergen concentration as a proxy for exposure, recent studies have revealed that indoor cockroach allergen exposure, but not mite or cat allergen exposure, is a significant risk factor for asthma [15]. Furthermore, despite the fact that the data collected on household characteristics varied greatly between the studies and that building materials and techniques are very different in different parts of the world, some common themes have emerged [20]. For example, in [16] mould concentrations varied hardly between four investigation areas, and neither climatologically conditions nor differences between urban and rural regions exhibited a systematic influence [16]. In another example [21], no association was found between the concentration of mite allergens and the environmental characteristics (geographic location, floor above ground, type of ventilation) and no correlation was found between indoor humidity and allergen levels [21].

The explanation comes if we consider that house dust mites live in an environment where there is no liquid water, and is dependent upon ambient humidity to absorb water from the atmosphere [20]. To get this water dust mites can obtain it by diffusion through the body or extract the water vapour from air via hygroscopic crystals in their supracoxal glands, located at the base of their first pair of legs [20]. The optimum RH for mite growth is 75-95%, at temperatures of 15-30 °C, while above 70% RH conditions may be optimal for fungal growth [17]. RH has a major influence on the survival of mite colonies and therefore levels of mite allergens.

Although laboratory and early field studies suggested that there was a strong relationship between RH and mite allergen levels, this had not been conformed by more recent large scale studies when other factors have been considered in a multivariate analysis. For example, freezing and /or dry weather can damage fungi and reduce the spore counts on outdoor samples, but the conditions indoors may be very hospitable to fungal growth non-seasonally [22].

From these studies we can conclude that outdoor RH influences indoor RH, but other household factors can influence mite allergen levels and that mean allergen levels in different geographical areas tend to be influenced by the local climate [20]. As a result of this, novel techniques have recently been developed which allows measurements of RH to be made within the mite microhabitat; that is, where it matters, in the depth of the carpet or mattress. This has revealed that the RH in the carpet may be higher than that in the room air and that with different types of construction, the differences between room RH and floor RH will vary. This suggests that the RH in the room air does not necessarily reflect the RH in the microhabitat of the mite, in the depth of the carpet pile [20].

In conclusion, it also calls, by the International Energy Agency in its Annex 41, the identification and characterization of different sources of pollution and humidity in indoor environments based on real data. All this is aimed at finding new methods of energy saving and improving the quality of indoor environments. This is one of the main functions of the Department of Energy of the University of A Coruña, whose activities have been reflected in the final chapters of this book. Data on actual indoor environments as varied as offices, museums, houses, ships... was sampled. The results, once verified by relevant international conferences, meetings and papers have served as the basis for subsequent processes of simulation.

REFERENCES

[1] Charles, K.E. Fanger's Thermal Comfort and Draught Models. 2003. IRC-RR-162. Http://irc.nrc-cnrc.gc.ca/ircpubs.[Accessed December 2010]

[2] McNall, Jr, P.E., Jaax, J., Rohles, F. H., Nevins, R. G. and Springer, W. Thermal comfort (and thermally neutral) conditions for three levels of activity. 1967. ASHRAE Transactions, 73.

[3] Nevins, R.G., Rohles, F. H., Springer, W. and Feyerherm, A. M. A temperature-humidity chart for thermal comfort of seated persons. 1966. ASHRAE Transactions, 72(1), 283.

[4] Fanger P.O. Thermal comfort. Analysis and applications in environmental engineering. 1970. McGrawHill. ISBN:0-07-019915-9

[5] ASHRAE Standard 55-2004-Thermal Environmental Conditions for Human Occupancy.2004. ASHRAE.

[6] Kerka, W.F., and C.M. Humphreys. Temperature and humidity effect on odor perception.1956. ASHRAE Trans. 61:531–552.

[7] Cain WS. Perception of odor intensity and the time-course of olfactory adaptation. 1974. ASHRAE Trans. 80:53–75.

[8] Woods, J.E. Ventilation, health and energy consumption: a status report. 1979. *ASHRAE Journal.* 23–27.

[9] Cain, W.S., Leaderer, B.P., Isseroff, R., Berglund, L.G., Huey, R.J., Lipsitt, E.D. and Perlman, D. Ventilation requirements in buildings – I. Control of occupancy odour and tobacco smoke odour. 1983. *Atmospheric Environment.* 17, 1183–1197.

[10] Berglund, L. and Cain, W.S. Perceived air quality and the thermal environment. 1989. In: Proceedings of IAQ '89: The Human Equation: *Health and Comfort*, San Diego, 93–99.

[11] Gunnarsen, L. and Fanger, P.O. Adaptation to indoor air pollution. 1992. *Environment International,* 18, 43–54.

[12] Knudsen, H.N., Kjaer, U.D. and Nielsen, P.A. Characterisation of emissions from building products: long term sensory evaluation, the impact of concentration and air velocity. 1996. In: Proceedings of Indoor Air '96, Nagoya, internacional Conference on Indoor Air Quality and Climate, 3, 551–556.

[13] L. Fang, G. Clausen and P. O. Fanger. Impact of Temperature and Humidity on the Perception of Indoor Air Quality. 1998. Indoor Air, 8, 80-90.

[14] Parat S., Perdrix A., Fricker-hidalgo H., Saude I., Grillot R. And Baconniers P. Multivariate analysis comparing microbial air content of an air-conditioned building and a naturally ventilated building over one year. 1997. *Atmos. Env.* 31. 3, 441-449.

[15] Liu A.H. Something Old, something New: Indoor endotoxin, allergens and asthma. Paedriatic Respiratory Reviews. 2004. 5 (Suppl A), S65-S71.

[16] Jovanovic S., Felder-Kennel A., Gabrio T., Kouros B., Link B., Maisner V., Piechotowski I., Schick K., Schrimpf M., Weidner U., Zöllner I. And Schwenk M. Indoor fungi levels in homes of children with and without allergy history.2004. *International Journal of Hygiene and env. Health.* 207, 369-378.

[17] Liao C.M., Luo W.C., Chen S.C., Chen J.W., Liang H.M. 2004.Temporal/seasonal variation of size-dependent airborne fungi indoor/outdoor relationship for a wind-induced naturally ventilated airspace. 2004. *Atmospheric environment 38*, 4415-4419.

[18] Hargreaves M., Parappukkaran S., Morawska L., Hitchins J., He C. and Gilbert D. A pilot investigation into associations between indoor airborne fungal and non-biological particle concentrations in residential

houses in Brisbane, Australia. 2003. *The Science of the Total Environment.* 312. 89-101.

[19] Lauener R.P. Primary prevention of allergies. Revue française d'allergologie et d'immunologie clinique. 2003. 43. 423-426.

[20] Editorial. Housing characteristics and mite allergen levels: to humidity and beyond. 2001. *Clinical and experimental allergy.* 31.803-805.

[21] Perfetti L., Ferrari M., Galdi E., Pozzi V., Cottica D., Grignani E., Minoia C. and Moscato. House dust mites, cat and cockroach allergens in indoor work-places (offices and archives). 2004. *Science of the Total Environment.* 328. 15-21.

[22] Zhou G., Whong W.Z., Ong T. and Chen B. Development of a fungus-specific PCR assay for detecting low-level fungi in a indoor environment. 2000. *Molecular and Cellular Probes.* 14. 339-348.

Chapter 2

INTRODUCTION TO THERMAL COMFORT

INTRODUCTION

The ISO 7730 Standard [1] defines thermal comfort as the mental condition that expresses satisfaction with the thermal environment. This definition can be easily understood but it is difficult to express in physical parameters. This means that the thermal comfort is a function of many parameters and not just one, such as temperature.

The study of thermal comfort can be divided into two; thermal comfort and local thermal comfort. The first is responsible for the study of environmental conditions that, on average, allow thermal neutrality and achieve the maximum energy savings. The local thermal comfort is responsible for the study of areas of indoor environments that present special conditions.

2.1. GENERAL THERMAL COMFORT

The main objective of heating, ventilation and air conditioning is to provide comfort to the occupants removing or adding heat and humidity of the occupied space. Similarly, the main objective of the study of thermal comfort conditions is generally able to determine the conditions for achieving human internal thermal neutrality with minimal power consumption. To do this, the need arises to study the human body's response to certain environmental conditions.

A comfortable environment is considered to be one where there is no thermal perturbation, namely that the individual does not feel too cold or hot. This is achieved when the brain interprets the signals as if it were the case of two opposing forces, with the sensations of cold working in one direction and heat on the other. If the signals received in both directions are the same magnitude the resulting feeling is neutral. A person in thermal neutrality and completely relaxed is in a special situation, where none of the cold or heat sensors are activated.

To define the thermal comfort conditions of a climate it must be given some characteristic parameters of the environment and its occupants. These parameters allow comparisons between the different environments of study. One should bear in mind that only after a thorough research of thermal comfort and indoor air quality can the quality of the thermal environment and, consequently, the efficiency of the HVAC systems be judged. Now, the most important parameters in the design of the facilities of the air-conditioning systems will be shown.

a) PMV scale

It is a computational model for the evaluation of generic comfort conditions and predicts its limits. The PMV scale is constituted by seven thermal sensation points ranging from -3 (cold) to +3 (hot), where 0 represents the neutral thermal sensation.

b) PPD scale

Even when the PMV index is 0, there are some individual cases of dissatisfaction with the temperature level, although all are dressed in a similar way and that the level of activity is the same. This is due to some differences of approach in the evaluation of thermal comfort from one person to another. To predict the number of people who are dissatisfied in a given thermal environment, an index called PPD is used. In this PPD index, individuals who vote -3, -2, -1, +1 +2 +3 on the PMV scale are considered thermally unsatisfied. Its evolution, as a function of PMV, is reflected in Figure 2.1.1.

Finally, ASHRAE proposed reference values between the PMV and PPD by outlining the assumptions on the rates of Fanger's PMV and PPD. This would indicate that for a PMV value between -0.85 and +0.85, the percentage of dissatisfaction (PPD) is 20% and the assumption of a stricter PPD of 10%, corresponds to a PMV between -0.5 and +0.5. The model defined by

Introduction to Thermal Comfort

ASHRAE presents a velocity limit of 0.2 m/s, since a greater air velocity allows higher temperatures, which can reach thermal comfort levels.

The ASHRAE standard specifies the conditions where a fraction of the occupants find the environment thermally acceptable. Manner, consistent with the ISO 7730 and 7726 [2] shows the new calculating methods of the PMV and PPD, does a classification of environments depending on the degree of thermal comfort and adds a new concept called adaptation, which can be used to assess facilities and indoor environments.

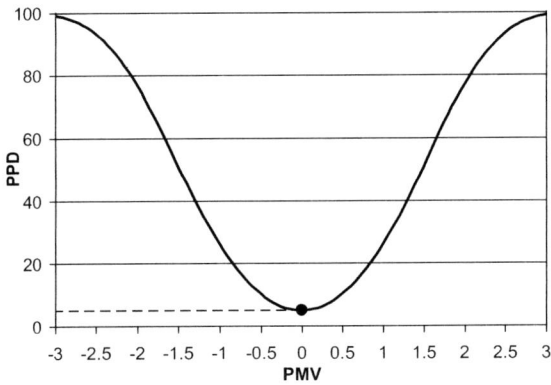

Figure 2.1.1. Evolution of PPD on the basis of PMV.

Table 2.1.1. Predicted percentage of dissatisfied (PPD) based on the predicted mean vote (PMV)

Comfort	PPD	Range del PMV
A	<6	-0.2 < PMV < 0.2
B	<10	-0.5 < PMV < 0.5
C	<15	-0.7 < PMV < 0.7

There can be three kinds of comfort zones, depending on the ranges PPD and PMV admissible as reflected in Table 2.1.1.

c) Estimated metabolic rate (met)

This is the amount of energy emitted by an individual as a function of the level of muscle activity. Traditionally, metabolism has been estimated at met

(1 met=58.15 W/m² surface of the body), and values for different activities are specified in the ISO 7730.

d) Estimated cloth insulation (clo)

Usually, the unit used to measure the insulation of clothing is produced by the clo, but the more technical unit and more frequently used is the m²°C/W (1 clo=155 m²°C/W). The scale is such that a naked person has a value of 0.0 clo and typical street clothes have 1.0 clo. Clo values, which correspond to different garments, are outlined in the ISO 7730. The value of the clo, for dressed people, can be calculated as an addendum to the clo of each garment.

e) Parameters that describe the thermal environment

To deduce the comfort equation, the comfortable temperature in the skin and the thermal balance equation of the body must be combined. This equation describes the average value of the relationship between measures of physical parameters and thermal sensation experienced by a neutral person.
In the sample of thermal conditions of an interior environment, it is important to remember that the human body does not feel the temperature of the compound; he feels the losses that occur with the thermal environment. Therefore, the parameters to be measured are those which affect the loss of heat and they are: air temperature (t_a), average temperature radiant ($\bar{t}_r$), air velocity (v) and absolute humidity of the air (w).

Mean Radiant Temperature

The mean radiant temperature $\bar{t}_r$ is defined as a uniform temperature in an imaginary black enclosure in which a person would experience the same losses by radiation than in the real compound. As the determination of $\bar{t}_r$ is difficult and laborious another parameter that can replace it is usually employed such as the operating temperature, t_o, which integrates the effect of $t_a + \bar{t}_r$, as we can see in Figure 2.1.2.

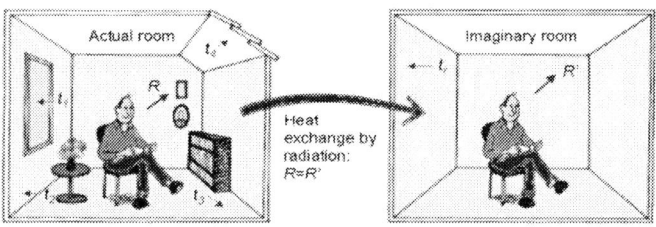

Figure 2.1.2. Mean radiant temperature (Courtesy of LumaSense Technologies).

Operating Temperature

The operating temperature, t_o, is the temperature at which must be the walls and the air of an equivalent compound that experiments the same heat transfer to the atmosphere by convection and radiation than in a enclosure where such temperatures are different. ASHRAE [3] shows some procedures for calculating the operating temperature. For example, this temperature can be calculated from the definition of the sensitive heat loss dissipated by the body, per unit of time and surface, as shown in the Eq. 2.1.1.

$$t_o = \frac{(h_r \bar{t}_r + h_c t_a)}{(h_r + h_c)} \qquad (2.1.1)$$

where;

t_a, $\bar{t}_r$ and t_o are the air, mean radiant and operative temperatures.

In other cases, for occupants engaged in near sedentary physical activity (with metabolic rates between 1.0 met and 1.3 met), not in direct sunlight, and not exposed to air velocities greater than 0.20 m/s the relationship can be approximated with acceptable accuracy by the Eq. 2.1.2.

$$t_o = \frac{(\bar{t}_r + t_a)}{2} \qquad (2.1.2)$$

This means radiant temperature can be calculated from the average temperature of the walls which is derived from the weighted average temperatures of each wall, through the Eq. 2.1.3. In this equation S represent the surfaces and t the temperatures.

$$t_{mp} = \sum_{i=1}^{i=n} S_i \frac{t_{pi}}{S_T} \quad (2.1.3)$$

Finally, it can define a comfort zone for some given values of humidity, air speed, metabolic rate and insulation produced by clothing, in terms of operating temperature or in terms of the combination of air temperature and the average radiant temperature. This area is represented in Figure 2.1.3 for air speeds not greater than 0.20 m/s. Two zones are usually shown by the ASHRAE, one for 0.5 and the other for 1.0 clo of insulation corresponding to the typical clothing worn when the outdoor environment is warm or cool respectively. The Eq. 2.1.4 and 2.1.5 define the operative temperature rage allowed for each case.

$$T_{op\,min,Icl} = [(I_{cl} - 0.5clo)T_{min\,1.0clo} + (1.0clo - I_{cl})T_{min\,0.5clo}]/0.5clo \quad (2.1.4)$$

$$T_{op\,max,Icl} = [(I_{cl} - 0.5clo)T_{max\,1.0clo} + (1.0clo - I_{cl})T_{max\,0.5clo}]/0.5clo \quad (2.1.5)$$

where;

$T_{max,Icl}$ is the upper operative temperature limit for clothing insulation I_{cl}.
$T_{min,Icl}$ is the lower operative temperature limit for clothing insulation I_{cl}.
I_{cl} is the thermal insulation of the clothing in question (clo).

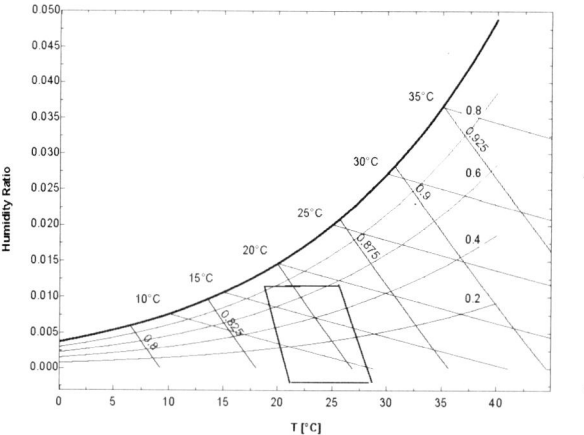

Figure 2.2. Comfort zone.

Relative Humidity

The relative humidity of the interior environment has often been considered of little importance in the design of air conditioning elements. But now it has become apparent effect on the comfort [4] and Wargocki in 1999 [5], the perception of indoor air quality [6], the health of the occupants [7] and energy consumption [8]. These concepts take on a greater importance if we bear in mind that in developed countries, the population uses 90% of their time indoors.

Air Velocity

A clear link between air velocity and thermal comfort has not yet been established. For this reason, the ASHRAE let an air speed rise to let a higher air temperature, but maintaining conditions within the comfort zone. In this way, a series of curves of allowed temperature for a given air speed, which are equivalent to those that produce the same heat loss through the skin can be found.

2.2. THERMAL COMFORT AND ALTERNATIVE THERMAL MODELS

To determine the thermal comfort rates of an environment, methods based on the study of the thermal balance of the human body (Fiala [9]) and empirical equations can be used. This last method employs equations that define the same comfort rates with greater simplicity than the first. Another of its advantages is that they are expressed in terms of parameters much more easily of sample in long periods of time and, therefore, will let relate the environment quality with energy savings. This section provides a detailed description of both methods. To analyse the sampled data we must do an automated calculation like that done by spreadsheets where the equations that relate the different thermodynamic variables are employed.

2.2.1. Ashrae Model

This model is shown by the ASHRAE [10] and, subsequently, it was implemented by Simonson [8] to determine the influence of coatings on indoor environments.

The basic equations that relate the temperature and relative humidity are well known and are based on the working assumptions of the moist air model. They believed the air was a mixture of ideal gases that could be defined as the dry air and water vapour. This water vapour presents an enthalpy equal to the enthalpy of the saturated steam at the same temperature, for a range of temperatures from -10 °C and 50 °C.

The relative humidity can be defined from the relationship between the partial pressure of water vapour in the air (p_v) and the partial pressure of water vapour in the saturated air (p_{vsat}). Therefore, the relative humidity can be expressed in Eq.2.2.1.1.

$$RH = \frac{P_V}{P_{VSAT}} \qquad (2.2.1.1.)$$

Furthermore, the partial water vapour pressure in the saturation conditions is a function of temperature (T) as is showed in Eq.2.2.2.

$$P_{vsat} = f(T) = e^F \qquad (2.2.1.2)$$

where the value of F is defined by the (ASHRAE).

27 K<T<273 K

$$F = \frac{C_1}{T} + C_2 + C_3 T + C_4 T_2 + C_5 T_3 + C_6 T_4 + C_7 \ln T$$

273 K<T< 473 K

$$F = \frac{C_8}{T} + C_9 + C_{10} T + C_{11} T^2 + C_{12} T^3 + C_{13} \ln T$$

The values of the constants are as follows:

$C_1 = -5674.5359$ $C_2 = 6.3925247$ $C_3 = -9.677843 \times 10^{-3}$
$C_4 = 6.22115701 \times 10^{-7}$ $C_5 = 2.0747825 \times 10^{-9}$ $C_6 = -9.484024 \times 10^{-13}$
$C_7 = 4.1635019$ $C_8 = -5800.2206$ $C_9 = 1.3914993$
$C_{10} = -4.8640239 \times 10^{-2}$ $C_{11} = 4.1764768 \times 10^{-5}$ $C_{12} = -1.4452093 \times 10^{-8}$
$C_{13} = 6.5459673$

From this equation it can be followed that when the temperature drops, the pressure of saturated vapour does so as well.

Another variable widely used is the absolute humidity (w) variable, which is defined as the ratio between vapour and the dry air mass. Absolute humidity is calculated from the relationship between the partial pressure of water vapour and the partial pressure of the air (p_a), as reflected in the Eq. 2.2.1.3.

$$w = 0.62198 \frac{p_v}{p_a} \qquad (2.2.1.3)$$

The equation above describes the relationship between temperature, humidity ratio and relative humidity, which can be expressed graphically through a psychrometric chart.

Finally, he ideal gas law shows that moist air enthalpy (h), represents the sum of the energy of its components (dry air and water vapour). If the temperature and humidity increases, the enthalpy of the air increases as is shown in Eq. 2.2.1.4 and represented in Figure 2.2.1.1.

$$h = c_{pa}t + w(c_{pw}t + L_o) \qquad (2.2.1.4)$$

This equation has been used in a simplified form by Simonson to how it is expressed in Eq. 2.2.1.5.

$$h = t + w(2501.6 + 1.865t) \qquad (2.2.1.5)$$

2.2.2. Equations for the Evaluation of Individual Thermal Balance

Thermal balance is totally accepted and followed by ISO 7730 [1] for the study of comfort conditions, regardless of the climatic region. Empirical equations have been sporadically used in specific climatic regions. The thermal balance begins with two necessary initial conditions to maintain thermal comfort:

1) A neutral thermal sensation must be obtained from the combination of skin temperature and full body temperature.

2) In a full body energy balance, the amount of heat produced by metabolism must be equal to that lost to the atmosphere (steady state).

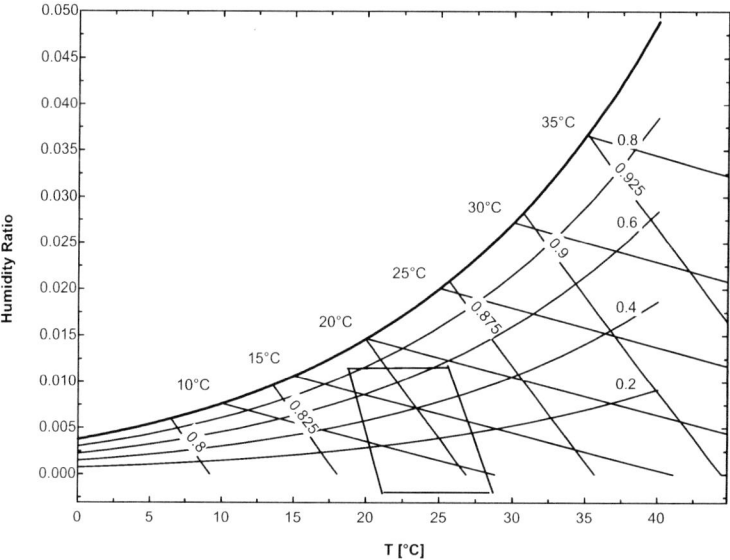

Figure 2.2.1.1. Psychometric chart.

Applying the above principles, Eq. 2.2.2.1. was obtained

$$M - W = q_{sk} + q_{res} + S \qquad (2.2.2.1)$$

$$M - W = (C + R + E_{sk}) + (C_{res} + E_{res}) + (S_{sk} + S_{cr}) \qquad (2.2.2.2)$$

where;

M rate of metabolic heat production (W/m^2)
W rate of mechanical work accomplished (W/m^2)
q_{sk} total rate of heat loss from skin (W/m^2)
q_{res} total rate of heat loss through respiration (W/m^2)

C+R sensible heat loss from skin (W/m²)
C_{res} rate of convective heat loss from respiration (W/m²)
E_{res} rate of evaporative heat loss from respiration (W/m²)
S_{sk} rate of heat storage in skin compartment (W/m²)
S_{cr} rate of heat storage in core compartment (W/m²)

The rate of heat storage in the body, considered as two nodes (skin and core) can be defined by the two equations 4.3 and 4.4.

$$Scr = \frac{(1-\alpha_{sk})mc_{p,b}}{A_D} \cdot \frac{dt_{cr}}{d\theta} \qquad (2.2.2.3)$$

$$Ssk = \frac{\alpha_{sk} mc_{p,b}}{A_D} \cdot \frac{dt_{sk}}{d\theta} \qquad (2.2.2.4)$$

where;

α_{sk} Fraction of body mass concentrated in skin
m body mass (kg)
$c_{p,b}$ specific heat capacity of body (kJ/kgK)
A_D DuBois surface area (m²)
t_{cr} temperature of core node (°C)
t_{sk} temperature of skin node (°C)
θ time (s)

The comfort equation can be obtained by setting the heat balance in thermally comfortable conditions for an individual, as the Eq. 2.2.2.1 shows. Based on these parameters the indices used in general to define a thermal environment can be established, as is shown in the equation 2.2.2.5, that predicts the mean vote and 2.2.2.6 of the percentage of dissatisfied.

$$PMV = (0.303 \cdot e^{-0.036 \cdot M} + 0.028) \cdot L \qquad (2.2.2.5)$$

$$PPD = 100 - 95 \cdot e^{-(0.03353 PMV^4 + 0.2179 PMV^2)} \qquad (2.2.2.6)$$

where L is the thermal load on the body, defined as the difference between internal heat production and heat loss to the actual environment.

The evaporative heat loss E_{sk} from the skin depends on the amount of moisture on the skin and the difference between the water vapour pressure on the skin and in the ambient environment as we can see in Eq. 2.2.2.7.

$$E_{sk} = \frac{w(p_{sk} - p_a)}{R_e + 1/(f_{cl}h_e)} \qquad (2.2.2.7)$$

where;

w is the skin wetted ness
p_{sk} is the water vapour pressure on skin (kPa)
Re is the evaporative heat transfer resistance of a layer of clothing (m^2 kPa)/W
He evaporative heat transfer coefficient (W/m^2 kPa)

The exchange of respiratory heat by convection and evaporative heat is shown in Eq. 2.2.2.8 and 2.2.2.9, respectively and Eq. 2.2.2.10 calculates sensible heat loss from the skin.

$$C_{res} = 0.0014 \cdot M \cdot (34 - t) \qquad (2.2.2.8)$$

$$E_{res} = 1.72 \cdot 10^{-5} \cdot M \cdot (5867 - P_v) \qquad (2.2.2.9)$$

$$C + R = \frac{(t_{sk} - t_o)}{R_{cl} + 1/(f_{cl}h)} \qquad (2.2.2.10)$$

where;
f_{cl} is the clothing area factor
R_{cl} is the thermal resistance of clothing (m^2 K)/W
t_{sk} is the temperature of the skin (°C)
h is the sum of convective and linear radioactive heat transfer coefficients (W/m^2 K)

Finally, in the case of office workers, external work W can be considered zero.

To deduce the comfort equation, the comfortable temperature of the skin and the sweat production equation is combined with the full body thermal balance [11]. This equation describes the relationship between measures of physical parameters and thermal sensation experienced by a person in an indoor environment. The comfort equation is an operational tool where physical parameters can be used to assess the thermal comfort conditions of an indoor environment. However, the comfort equation obtained by Fanger [4] is too complicated to be solved through manual procedures.

After studying the equations that define the heat balance of a person we can deduce the need of sample the instantaneous evolution of operative temperature, air velocity and relative humidity. To collect the thermal comfort data, we can employ transducers as that employed by the thermal comfort module of Innova Airtech 1221 [12], that will be showed in chapter 4. To facilitate the understanding of this procedure the parameters that must be measured directly or calculated are summarised and indicated in Table 2.2.2.1.

Table 2.2.2.1. Methods to calculate general thermal comfort indexes

	Air velocity (v_a)	Air temperature (t_a)	Mean radiant temperature ($\bar{t}_r$)	Humidity (w)
Method 1	Measure	Measure	Calculate	Measure
Method 2	Air velocity (v_a)	Operative temperature (t_o)		Humidity (w)
	Measure	Measure		Measure
Method 3	Equivalent temperature (t_{eq})			Humidity (w)
	Measure			Measure
Method 4	Air velocity (v_a)	Effective temperature (ET*)		
	Measure	Calculate		

In Table 2.2.2.1 we find the term "Equivalent Temperature", which is often used instead of Dry Heat Loss. This equivalent temperature can be calculated from the dry heat loss and, by definition, is the uniform temperature of a radiant black enclosure with zero air velocity in which an occupant would have the same dry heat loss as the actual non-uniform environment.

2.2.3. Alternative Models

2.2.3.1. Alternative PMV Models

Among all the thermal environment indices, the principal one is the PMV. The work done by Oseland, and subsequently reflected by the ASHRAE, concluded that the PMV can be used to predict the neutral temperature with a margin of error of 1.4 °C compared to the neutral temperature defined by the equation of thermal sensation. This thermal sensation expresses an equivalent index to the PMV. Its principal difference is that thermal sensation is obtained by regression of a surveys to different individuals located in an environment. This survey presents a scale as that shown in Table 2.2.3.1.1.

Table 2.2.3.1.1. Thermal sensation values

Tsens	Thermal sensation
3	Warm
2	Heat
1	soft
0	Neutral
-1	Soft freshness
-2	Freshness
-3	Cold

An example of a thermal sensation model that takes into account the effect of the clo, has been developed by Berglund [13] and is shown in Eq. 2.2.3.1.1.

$$T_{sens} = 0.305 \cdot T + 0.996 \cdot clo - 8.08 \qquad (Eq.\ 2.2.3.1.1)$$

Brager [14] also showed that the PMV was found to be lower (colder) than the obtained thermal sensation when he studied office buildings in San Francisco during the winter. So with that he defined a neutral temperature of 24.8°C which was 2.4°C above the estimated value. After considering various UK offices with mechanical ventilation it was demonstrated that the PMV differed by 0.5 points with the thermal sensation, which is equivalent to say 1.5°C differences.

In Australia, De Dear and Auliciems (1985) [15], found a difference of 0.5°C to 3.2 °C between the neutral temperatures, estimated by surveys, and determined by the PMV model of Fanger. Subsequently, Dear conducted a study in twelve Australian office buildings, and a temperature difference was defined between the neutral temperatures proposed by surveys and by the PMV determined in about 1 °C. Dear et al. extended his studies to those made by Brager. In this way, they have returned to analyze and correct the data by the seat isolation. Again, it was found discrepancies between the neutral temperature, based on the value obtained through surveys, and the value predicted by the equations. As a result, it appears that for real conditions, the thermal sensation of neutrality is in line with a deviation of the order of 0.2 °C to 3.3 °C, and an average of 1.4 °C of thermal neutrality conditions. The error was attributed to either a PMV erroneous definition of the metabolic activity and the index of clo, or to being unable to take into account the isolation of the seat.

The Institute for Environmental Research at Kansas State University, under ASHRAE contract, has conducted extensive research on the subject of thermal comfort in sedentary regime. The purpose of this investigation was to obtain a model to express the PMV in terms of parameters easily sampled in an environment.

In consequence, an investigation of 1600 school-age students revealed statistics correlations between the level of comfort, temperature, humidity, gender and exposure duration. Groups of five men and five women were exposed to a range of temperatures between 15.6 °C and 36.7 °C with increases of 1.1 °C at eight different relative humidities of 15, 25, 34, 45, 55, 65, 75 and 85% and for an air speeds of lower than 0.17 m/s. During a study period of three hours with intervals of half an hour, subjects reported their thermal sensations on a ballot paper with seven categories ranging between -3 and 3, as reflected in Table 2.2.3.1.1. These categories show a thermal sensation that varies between cold to warm, passing through 0 that indicates thermal neutrality.

The results have yielded to an expression of the form shown in Eq. 2.2.3.1.2.

$$PMV = a \cdot t + b \cdot p_v - c \qquad (2.2.3.1.3)$$

By using this equation and taking into account gender and exposure time to the indoor environment, constants as those displayed in Table 2.2.3.1.2. should be used. With these criteria it has been given a comfort zone that, on average, is close to conditions of 26 °C and 50% relative humidity. They have undergone the study subjects to a sedentary metabolic activity, dressed with normal clothes and with a thermal resistance of approximately 0.6 clo. Their exposure to the indoor ambiences was of three hours.

2.2.3.2. Operative Temperature Model

According to the ASHRAE, temperature can be obtained on the basis of activity and clothing as is shown in the Eq. 2.2.3.2.1.

$$t_{oac} = t_{osed} - 3(1 + clo)(met - 1.2) \qquad (2.2.3.2.1)$$

The equation is valid between 1.2 and 3 met with a minimum t_{oac} acceptable value of 15°C.

Table 2.2.3.1.2. The coefficients a, b and c are a function of spent time and the sex of the subject

Time/Sex	A	B	C
1 Hours/Man	0.220	0.233	5.673
Woman	0.272	0.248	7.245
Both	0.245	0.248	6.475
2 Hours/Man	0.221	0.270	6.024
Woman	0.283	0.210	7.694
Both	0.252	0.240	6.859
3 Hours/Man	0.212	0.293	5.949
Woman	0.275	0.255	8.620
Both	0.243	0.278	6.802

The operative temperature conditions for a sedentary activity in summer or winter are:

Terms of summer: $\qquad t_{osed} = 24.5 \pm 1.6°C \qquad (2.2.3.2.2)$

Terms of winter: $\qquad t_{osed} = 21.8 \pm 1.8°C \qquad (2.2.3.2.3)$

The ASHRAE recommends an office building activity level between 1.1 and 1.3 met. We must remember that a met is about 58.2 W/m^2. ASHRAE also provides the clo value for clothing appropriate to each season. In this way, typical Spanish clo values chosen for each season of the year were 0.5 in summer and 1.0 in winter.

2.2.3.3. Adaptive Models

Over the last few years, adaptive models have been applied to define the neutral temperature as a function of outdoor, indoor or both temperatures. Some of them present a higher accuracy in certain conditions. Nicol and Roaf [16] recommended the Equation 2.2.3.3.1 model for occupants of naturally ventilated buildings. Many other adaptive models have also been proposed. For example, Humphreys [17] developed two models for neutral temperature, as given in Equation 2.2.3.3.2 and 2.2.3.3.3, and Auliciems and de Dear developed the relations for predicting group neutralities based on mean indoor and outdoor temperatures, as shown in Equations 2.2.3.3.4, 2.2.3.3.5 and 2.2.3.3.6, which was employed by the ASHRAE in Equation 2.2.3.3.7.

$$T_{n,o} = 17 + 0.38 T_o \quad (2.2.3.3.1)$$

$$T_{n,1} = 2.6 + 0.831 T_i \quad (2.2.3.3.2)$$

$$T_{n,o} = 11.9 + 0.534 T_o \quad (2.2.3.3.3)$$

$$T_{n,i} = 5.41 + 0.731 T_i \quad (2.2.3.3.4)$$

$$T_{n,o} = 17.6 + 0.31 T_o \quad (2.2.3.3.5)$$

$$T_{n,i,o} = 9.22 + 0.48 T_i + 0.14 T_o \quad (2.2.3.3.6)$$

ASHRAE:

$$T_c = 17.8 + 0.31 T_o \quad (2.2.3.3.7)$$

where T_c is the comfort temperature, T_o is the outdoor air temperature, T_i is the mean indoor air temperature, $T_{n,i}$ is neutral temperature based on mean indoor air temperature and $T_{n,o}$ is neutral temperature based on mean outdoor air temperature.

Before applying these models, we must remember that occupants must be engaged in near sedentary activity (1-1.3 met) and must be able to freely adapt their clothing. Furthermore, neither a heating system nor a mechanical cooling system can be in operation, although non-conditioned mechanical ventilation can be present. Despite this, windows must be the principal way of controlling the thermal conditions.

2.3. LOCAL THERMAL COMFORT

Once certain thermal conditions that, on average, affect the environment since (general thermal comfort) and, in order to ensure thermal comfort from its broader view, there arises the need to focus on the study of areas subject to special conditions (local thermal comfort). In these areas it is possible to improve comfort through the relevant structural modifications, its own air conditioning facilities and people's habits.

2.3.1. Air Velocity Models

Air velocity affects sensible heat dissipated by convection and latent heat dissipated by evaporation, since both the convection coefficient and the amount of evaporated water per unit of time depend on it, and therefore the feeling of comfortability becomes affected by air draughts. Aiming towards energy saving in summer, the ambient air temperature can be kept slightly higher than the optimum and achieve a more pleasant by increasing air velocity. The maximum acceptable air speed is of 0.9 m/s.

In winter, the air circulation causes a cold feeling and forces to keep air temperature above that needed to avoid a feeling of discomfort, with its corresponding energy consumption. Bearing in mind that in this season, the dry air temperature tends to be in the low band of comfort, air conditions in inhabited areas must be carefully studied, in order to maintain the conditions of wellbeing without wasting energy. It is recommended that the winter air velocity in the occupied zone should be lower than 0.15 m/s. Localized draft problems are more common in indoor environments with air conditioning, vehicles and aircraft. Even without a speed-sensitive air, there may be dissatisfaction because of excessive cooling somewhere in the body. In principle, there is utmost sensitivity to currents on the nude parts of the body. Therefore, usually only noticeable flows on the face, hands and lower legs. The amount of heat lost through the skin because of the flow depends on the average speed of air, temperature and turbulence in the flow.

Due to the behaviour of the cold sensors on the skin, the degree of the sense of discomfort depends not only on the losses of local heat, but also on the influence in temperature fluctuations. For equal thermal losses, there is greater sense of dissatisfaction with high turbulence in the air flow. There are some studies on the types of fluctuations that cause greater dissatisfaction. These have been obtained from groups of individuals subjected to various air speed frequencies. The oscillations with a frequency of 0.5 Hz are the most uncomfortable, while oscillations with a higher frequency of 2 Hz produce some effects much less sensitive effects. According to the ISO 7730 [1], drafts produce an unwanted local cooling in the human body. The flow risk can be expressed as the percentage of annoyed individuals and be calculated by Eq. 2.3.1.1.

$$DR = (34 - t)(v - 0.05)^{0.62}(0.37vT_u + 3.14) \qquad (2.3.1.1)$$

The draught risk model is based on studies of 150 subjects exposed to air temperatures between 20-26°C, with average air speed between 0.05-0.4 m/s and turbulence intensities from 0-70%. The model is also applicable to low densities of people, with sedentary activity and a neutral thermal sensation over the full body. The draught risk is lower for non- sedentary activities and for people with neutral thermal sensation conditions. Figure 2.3.1.1 shows this relationship between air speed, temperature and the degree of turbulence, for a percentage of dissatisfaction of 10 or 20%. The different curves refer to a percentage of turbulence from 10 to 80%.

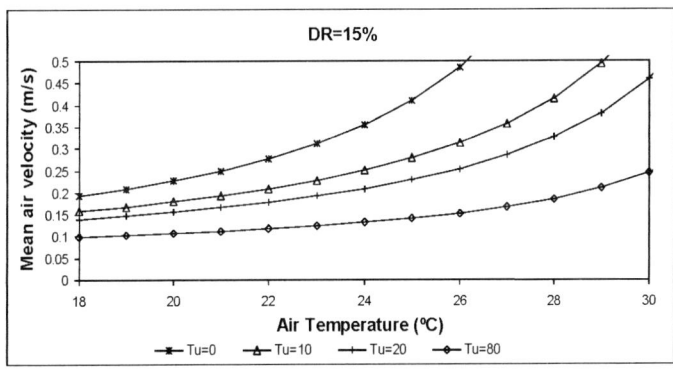

Figure 2.3.1.1. Average air velocity, depending on temperature and the degree of turbulence thermal environments for type A, B and C.

2.3.2. Asymmetric Thermal Radiation

A person located in front of an intense external heat source in cold weather may notice, after a certain period of time, some local dissatisfaction. The reason is the excessive warm front and high cooling on the opposed side. The uncomfortable situation could be remedied with the frequent change of position to achieve a more uniform heating. This example shows the uncomfortable conditions due to a non uniform radiant heat effect.

To evaluate the non uniform thermal radiation the asymmetric thermal radiation parameter ($\bar{t}_r$) is used. This parameter is defined on the basis of the difference between the flat radiation temperature (t_{pr}) of the two opposite sides of a small plane element. The experiences with individuals exposed to variations in asymmetrical radiant temperature, such as are the conditions caused by warm roofs and cold windows, produce the greatest impact of dissatisfaction. During previous experiences all surfaces of the enclosure and air temperature were preserved.

The Parameter can be obtained by two methods; the first method is based on the measure in two opposite directions and using a transducer to capture radiation that affects a small plane from the corresponding hemisphere. The other method is to obtain temperature measurements from all surfaces of your surroundings and calculating the Δt_{pr} Eq. 2.3.2.1 -2.3.2.4 show the employed models for each case.

a) Hot ceiling ($\Delta t_{pr} < 23°C$)

$$PD = \frac{100}{1 + \exp(2.84 - 0.174 \cdot \Delta t_{pr})} - 5.5 \qquad \text{(Eq. 2.3.2.1)}$$

b) Cold wall ($\Delta t_{pr} < 15°C$)

$$PD = \frac{100}{1 + \exp(6.61 - 0.345 \cdot \Delta t_{pr})} \qquad \text{(Eq. 2.3.2.2)}$$

c) Cold Ceiling ($\Delta t_{pr} < 15°C$)

$$PD = \frac{100}{1 + \exp(9.93 - 0.50 \cdot \Delta t_{pr})} \qquad \text{(Eq. 2.3.2.3)}$$

d) Hot wall ($\Delta t_{pr} < 35°C$)

$$PD = \frac{100}{1+\exp(3.72 - 0.052 \cdot \Delta t_{pr})} - 3.5 \qquad (Eq.\ 2.3.2.4)$$

Finally, the curves obtained are reflected in Figure 2.3.2.1.

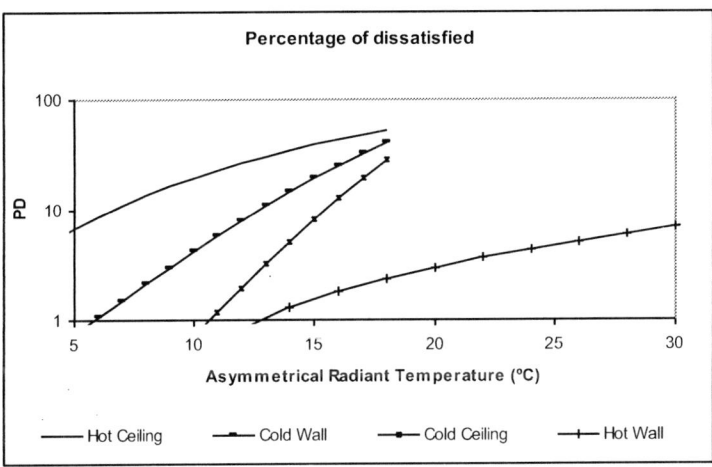

Figure 2.3.2.1. Percentage of dissatisfied as a function of asymmetrical radiant temperature, produced by a roof or wall cold or hot.

2.3.3. Vertical Temperature Difference

In general, there is an unsatisfied sensation with heat around the head cold in the feet, regardless of whether the cause is convection or radiation. We can express the vertical temperature difference of the air existing at the ankle and neck height, respectively.

Experiments on people's neutral thermal conditions have been conducted. Based on the results, it has been observed that a temperature difference between head and feet of 3 °C produces a dissatisfaction of 5%. The curve obtained is reflected in Figure 2.3.3.1. For a person sitting in a sedentary activity the ISO 7730 is accepted as the acceptable value of 3 °C. The corresponding model is shown in the Eq. 2.3.3.1.

$$PD = \frac{100}{1 + \exp(5.76 - 0.856 \cdot \Delta t\ )} \qquad \text{(Eq. 2.3.1.1)}$$

where;

Δt is the vertical temperature difference (°C)

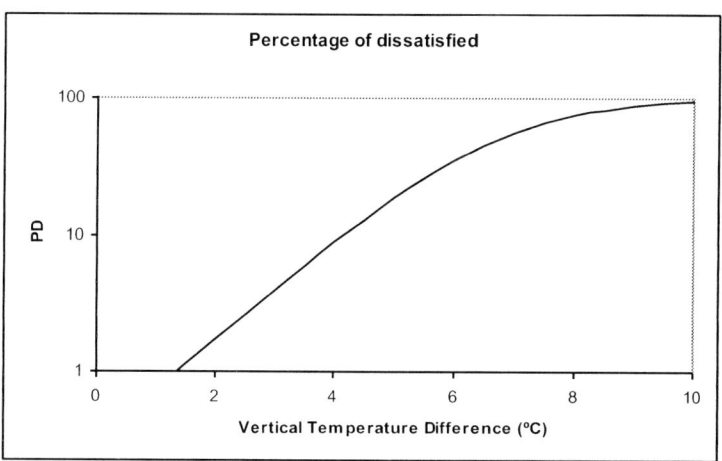

Figure 2.3.1.1. Percentage of dissatisfied, depending on the vertical temperature difference.

2.3.4. Soil Temperature

Local dissatisfaction due to the direct contact between the feet and the ground may be related with a temperature which is either too high or low. Heat losses are dependent on other parameters such as conductivity, the heat capacity of the ground material and insulation capacity of the entire foot-footwear. The ISO 7730 standard provides levels of comfort in sedentary activities for a 10% dissatisfied. This leads to the consideration of acceptable ground temperatures between 19 and 29 °C. Studies have allowed to obtain the curve shown in Figure 2.3.4.1 and Eq. 2.3.4.1 reflects the model of the percentage of dissatisfaction for different floor temperatures.

$$PD = 100 - 94 \cdot \exp(-1.387 + 0.118 \cdot t_f - 0.0025 \cdot t_f^2) \qquad \text{(Eq. 2.3.4.1)}$$

2.3.5. Local Thermal and Humidity Comfort Mathematical Models

For an indoor air quality study there are a number of empirical equations used by some authors over the last few years such as Simonson. These indices are employed to determine indices such as the percentage of dissatisfaction with local thermal comfort, thermal sensation and indoor air acceptability in terms of some simple parameter measures such as dry bulb temperature and relative humidity.

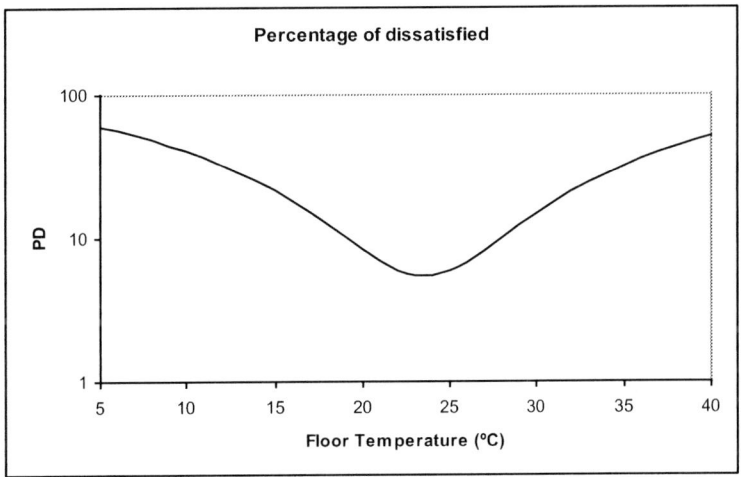

Figure 2.3.4.1. Percentage of dissatisfied, depending on the temperature of the floor.

In this sense, the humidity ratio and relative humidity are the most important parameters to compare the effect of moisture in the environment, while temperature and enthalpy reflect the thermal energy of each psychometric process.

Simonson showed that moisture had a small effect on thermal comfort, but a lot more on the local thermal comfort. The current regulations (ISO 7730, ASHRAE and DIN 1946) do not coincide in the exact value of moisture in the environment for some conditions, but conclude that a very high or very low relative humidity worsen comfort conditions.

The agreement chosen by the ANSI/ASHRAE and ISO 7730 to establish the comfort boundary conditions was about 10% of dissatisfaction. Other authors believe that the local thermal comfort is primarily a function of thermal gradient at different altitudes and air speeds, but may also be due to the presence of sweat on the skin or inadequate mucous membrane refrigeration.

To meet the local thermal comfort produced by the inside air conditions, Toftum et al. (1998b) [18-19] have studied the response of 38 individuals, who were provided with clean air in a closed environment. The air temperature conditions ranged between 20-29 °C and the humidity ratio between 6-19 g./kg. as from 20 °C and 45% RH to 29 °C and 70% RH. Individuals assessed the ambient air with 3 or 4 puffs and thus the equation for the percentage of local dissatisfaction was developed as shown in Eq. 2.3.5.1.

$$PD = \frac{100}{1 + e^{(-3.58 + 0.18 \cdot (30 - t) + 0.14 \cdot (42.5 - 0.01 p_v))}} \qquad (2.3.5.1)$$

The ASHRAE recommends keeping the percentage of local dissatisfaction below 15% and the percentage of general thermal comfort dissatisfaction below 10%. This PD tends to decrease when the temperature decreases.

2.4. HEAT STRESS AND WORK RISK PREVENTION

Very intense heat on the human body may cause aggressions. In consequence, in extreme situations it is necessary to strictly limit time in such conditions. In industry this restriction is implemented, in most cases, allowing workers free periods of activity and rest. This method usually leads to quite satisfactory results and involves a substantial risk that, in certain circumstances, workers extend their exposure to dangerous limits.

The method presented in the NTP Spanish Standard [21] let us calculate the maximum time that a worker can stay in a certain thermally aggressive situation and the length of time of the mandatory rest period following exposure is before they can recommence work. This method was published and presented in 1966 by McKarns and Brief [22] and is an adaptation of the so-called Heat Stress Index developed by Belding and Hatch [23] based on previous work by Haines and Hatch [24].

The method is based on the calculation of the magnitude of heat exchange between man and environment by three key mechanisms through which this exchange takes place: convection, radiation and evaporation.

The calculation is based on three key assumptions: men's standard weight of 70 kg, clothing is light (summer shirt and trousers or similar) and skin temperature is 35 °C. The temperature of the skin should not be confused with the internal temperature of the body that is what we feel about where we put the thermometer.

After calculating the magnitude of the exchanges that take place by convection and radiation, and the maximum amount of heat that the subject is able to eliminate through sweat evaporation (evaporation maximum Emax) in environmental conditions, the appropriate method for calculating the amount of heat that the individual should remove through evaporation to reach thermal equilibrium (gain = loss) is shown in Eq. 2.4.1.

$$E_{req} = M + R + C \qquad (2.4.1)$$

where:
Ereq = evaporation needed for balance (W)
M = heat generated by the body (metabolism) (W)
C = heat gained or lost by convection (W)
R = heat gained by radiation (W)

The difference between Ereq and Emax is clearly the highest net gain of body heat that is exposed to the heat subject. Acknowledging that exposure should cease when the body's internal temperature has risen 1 °C and that this increase is due to the fact that evaporation is less than the maximum needed for the thermal balance, the time needed to produce such an increase is given by the Eq. 2.4.2.

$$t_{ex} = \frac{3600}{(Ereq - E\max)} \qquad (2.4.2)$$

where;
t_{ex} is the maximum time spent in the environment under consideration, expressed in minutes.

By the same reasoning it is possible to calculate the required rest time between two successive exposures in the areas of rest which would allow the body to remove the heat accumulated during exposure to retrieve the internal temperature.

2.5. ACTUAL RELATED ISO STANDARD

Finally, the principle ISO, NTP and ASHRAE [10] principles that may be of interest to the ergonomic and thermal comfort investigator will be enumerated

a) ISO Standard

ISO 11399:1995. Ergonomics of the thermal environment-Principles and application of relevant International Standards.

ISO 7730:2005. Ergonomics of the thermal environment-Analytical determination and interpretation of thermal comfort using calculation of the PMV and PPD indices and local thermal comfort criteria.

ISO 9920:2007 Ergonomics of the thermal environment-Estimation of thermal insulation and water vapour resistance of a clothing ensemble.

ISO 8996:2004 Ergonomics of the thermal environment-Determination of metabolic rate.

ISO 7726:1998 Ergonomics of the thermal environment-Instruments for measuring physical quantities.

ISO 7933:2004 Ergonomics of the thermal environment-Analytical determination and interpretation of heat stress using calculation of the predicted heat strain

ISO 11079:2007 Ergonomics of the thermal environment-Determination and interpretation of cold stress when using required clothing insulation (IREQ) and local cooling effects.

ISO 9886:2004 Ergonomics-Evaluation of thermal strain by physiological measurements.

ISO 10551:1995 Ergonomics of the thermal environment-Assessment of the influence of the thermal environment using subjective judgement scales.

b) ASHRAE Standard

ANSI/ASHRAE 55-2004 Thermal environmental conditions for human occupancy.

c) NTP Spanish Standard

Spanish Thermal Comfort Standards
NTP 74: Thermal comfort.
NTP 242: Ergonomics: Ergonomic office workplaces analysis
NTP 501: Thermal environment: Local thermal discomfort

NTP 503: Acoustic comfort: noise in offices
NTP 358: Odours: A factor of indoor air quality and comfort
Spanish Work Risk Prevention Standards
NTP 279: Thermal environmental and dehydration.
NTP 322: Estimation of the heat stress: WBGT.
NTP 350: Heat stress evaluation. Required sweating index.
NTP 462: Cold stress: Occupational exposures evaluation.
NTP 18: Heat stress evaluation of severs exposures.
NTP 241: Controls and signals: perception ergonomics.
NTP 413: Workload and course of pregnancy.
NTP 534: Mental workload: factors.
NTP 445: Mental workload: fatigue.
NTP 575: Mental workload: indicators.
NTP 179: Mental Workload: definition and measurement.
NTP 226: Ergonomic design and accessibility of controls.
NTP 275: Mental workload in health care works: An assessment checklist.
NTP 575: Mental workload: indicators.
NTP 177: Physical work load: definition and measurement.
NTP 295: Physical work load evaluation by continuous register of heart rate.
NTP 405: Human factor and accident rates. Social aspects.
NTP 413: Workload and course of pregnancy.
NTP 445: Mental workload: fatigue.
NTP 387: Working conditions analysis: the ergonomic workplace analysis.

REFERENCES

[1] ISO 7730:2005. Ergonomics of the thermal environment-Analytical determination and interpretation of thermal comfort using calculation of the PMV and PPD indices and local thermal comfort criteria. 2005.
[2] ISO 7726: 2002. Ergonomics of the thermal environment - Instruments for measuring physical quantities. 2002.
[3] ASHRAE 55-2004 ASHRAE Standard 55-2004-Thermal Environmental Conditions for Human Occupancy.2004.
[4] Fanger, P.O. Thermal Comfort. Copenhagen: Danish Technical. 1970. Doctoral Thesis.

[5] Wargocki, P., Wyon, D.P., Baik, Y.K., Clausen, G. and Fanger, P.O., Perceived air quality, Sick Building Syndrome (SBS) symptoms and productivity in an office with two different pollution loads. 1999. Indoor Air, 9, 165–179.

[6] Fang L., Clausen G., Fanger P. O. Impact of Temperature and Humidity on Perception of Indoor Air Quality During Immediate and Longer Whole-Body Exposures. 1998. Indoor Air. Vol. 8, 4, 276-284.

[7] Molina M. Impacto de la temperatura y la humedad sobre la salud y el confort térmico, climatización de ambientes interiores. 2000. (Tesis doctoral) . Universidad de A Coruña.

[8] Simonson C. J., Salonvaara M. and Ojanen T. Improving Indoor Climate and Comfort with Wooden Structures. 2001. Technical research centre of Finland. Espoo.

[9] Fiala D., Lomas K.J., and Stohrer M. Computer prediction of human thermoregulatory and temperature responses to a wide range of environmental conditions. 2001. *Int. J. Biometeorol.*, 45, 143-159.

[10] ASHRAE Handbook - Fundamentals. 2005.

[11] Stanton, N. Brookhuis K., Hedge A., Salas E., Hendrick H. W. Handbook of Human Factors and Ergonomics Methods. CRC Press, 2005. ISBN 0415287006, 9780415287005.

[12] INNOVA Air Tech Instruments A/S (1997). Thermal Comfort [online]. Denmark. Available from: http://www.innova.dk/books/ thermal/ [Accesed 1 April 2007]

[13] Berglund, L. Mathematical Models for Predicting the Thermal Comfort Response of Building Occupant. 1978. ASHRAE Trans.84.

[14] Brager G.S, de Dear R.J. Thermal adaptation in the built environment: a literature review.1998. *Energy and buildings.* 27, 83-96.

[15] De Dear RJ, Auliciems A. Validation of the predicted mean vote model of thermal comfort in six Australian field studies.1985. *ASHRAE transactions.* 91 (2B), 452-68.

[16] Nicol F, Roaf S. Pioneering new indoor temperature standard: the Pakistan project.1996. *Energy and Buildings.* 23, 169-74.

[17] Humphreys MA. Comfortable indoor temperatures related to the outdoor air temperature. *Building Service Engineer* 1976; 44: 5-27.

[18] Toftum J., Jorgensen A. S., Fanger P.O. Upper limits for indoor air humidity to avoid uncomfortably humid skin.1998. *Energy and Buildings.* 28, 1-13.

[19] Toftum J., Jorgensen A.S., Fanger P.O. Upper limits of air humidity for preventing warm respiratory discomfort. *1998. Energy and Buildings* 28, 15-23.

[20] Normativa Técnica de Prevención (NTP). Ministerio de Industria. http://www.insht.es/portal/site/Insht/menuitem.a82abc159115c8090128ca10060961ca/?vgnextoid=db2c46a815c83110VgnVCM100000dc0ca8c0RCRD

[21] NTP 18: Estrés térmico. Evaluación de las exposiciones muy intensas. Normativa Técnica de Prevención (NTP). Ministerio de Industria. http://www.insht.es/

[22] Mckarns, J.S., Brief, R.S. Nomographs Give Refined Estimate of Heat Stress Index. 1966. Heating, Piping, Air Cond. 1, 113-116.

[23] Beldings, H.S., Hatch, T. F. Index for Evaluating Heat Stress in Terms of Resulting Physiological Strains. *Heating, Piping,1955. Air Cond.,* 8, 129-136.

[24] Haines, G.F., Hatch, T.F. Industrial Heat Exposures - Evaluation and Control. 1952. *Heating and Ventilating*, 49, 11, 93-104.

Chapter 3

INTRODUCTION TO INDOOR AIR QUALITY

INTRODUCTION

It is now accepted that those environments that do not have natural ventilation and are closed to achieve a better performance of the air conditioning system may be areas of exposure to pollutants. These include offices, public buildings, schools and nurseries, commercial buildings and even private residences. The extent of damage that this may pose to health is not exactly known, as the levels of pollutants that have been identified, mainly in studies in offices and homes, often far below the permissible limits of respective exposure to industrial environments.

The air quality inside the building is a function of a set of parameters that include air quality outside the partitioning, the design of air conditioning systems, the conditions in which this system works and is reviewed and the presence of pollutant sources and their magnitude. Obviously, the air inside a building must not contain contaminants in concentrations higher than those that can harm health or cause discomfort to the occupants. These pollutants include those that may be present in the outside air that enters the building and caused by the interior, furniture, construction materials, surface coatings and treatments of air. The most frequent risk to the occupants are: exposure to toxic substances, radioactive and irritant induction of infections or allergies, temperature and relative humidity conditions and odour nuisance.

The symptoms that are related to poor air quality inside a building are: headache, dizziness, nausea, fatigue, dry skin, eye irritation, sinus congestion and cough. It is often difficult to differentiate those caused directly between the environment and the psychological origin. Do not forget that poor air quality causes discomfort and can trigger complex psychological reactions, mood swings and difficulties in interpersonal relationships.

In recent years the European Collaborative Action has studied and identified the problems caused by the lack of quality of indoor environments under the heading of sick building syndrome (SBS) and tight building syndrome. This phenomenon has been recognized by the World Health Organization (WHO) [1] since 1986 and its objectives have been to identify their causes and eliminate them [2-9]. The techniques to neutralize or eliminate air pollutants have considerable importance and interest. The Principal techniques are:

3.1. Control of emission sources.
3.2. Renovation of indoor air.

These techniques will therefore be discussed in this chapter.

3.1. POLLUTION SOURCES

As many materials can contribute to the deterioration of indoor air quality, those that generate lower emissions and reduce pollution sources in ventilation systems will be employed. When the effect produced by polluting material is low, ventilation needs are reduced. The identification of the pollution load can be established from a chemical or sensory impairment, although there are other methods based on biological effects.

Chemical loading is usually expressed in terms of the release of chemicals from a source in $\mu g/s$ o ($\mu g/m^2 s$). To determine the total chemical load, we must summarise the estimate load contributed by each individual product to the air of an enclosed space. The sensory load can be quantified by the olf unit, which includes the effect of many chemicals, as they are perceived by humans. The occupants of an enclosed space emit biofluentes and possibly smoke snuff. Olf is defined as pollution produced by a healthy adult who does light office work and is in a neutral thermal environment. A standard (non-smoker) produces 1 olf, while an average smoker produces 6 olf. Table 3.1 shows the pollution load, of adult occupants who develop different activities, which include non-smokers and smokers in different percentages of the total number of occupants. There is also the pollution load produced by children. In addition, Table 3.1 also shows the human production of carbon dioxide, carbon monoxide and water vapour.

Table 3.1.1. Pollution caused by the occupiers

	Sensory pollution loads olf/person	CO_2, CO, H_2O l/(h.p)
Sedentary, 1-1.2 met		
0 % Smokers	1	19 50
20 % Smokers	2	19 11x10-3 50
40 % Smokers	3	19 21x10-3 50
100 % Smokers	6	19 53x10-3 50
Physical exercise		
Low level, 3 met	4	200
Half a level, 6 met	10	430
High level (athletes), 10 met	20	750
Children		
Preschoolers, 3-6 years, 2.7 met	1.2	90
School, 14-16 years, 1-1.2 met	1.3	19 50

3.1.1. Chemical and Biological Load

The occupants of a building are themselves a source of pollution, since human beings naturally carbon dioxide, water vapor, particles and biological aerosols. Another group has its origins in combustions that occur in the interior such as the use of cleaning products, used to maintain and beautify, which causes the presence of contaminants in the building. Some of these sources produce complex mixtures, such as snuff smoke, fumes and aerosols generated in the preparation of meals, allergens and infectious biological aerosols generated in the cooling circuits and themselves in the human body.

Although the problem is difficult to address a ranking of pollutants will be systemized in the next sections.

3.1.1.1. Chemical Pollutants

Table 3.1.1.1.1 presents the most common chemical contaminants in the air inside buildings classified according to their possible source.

3.1.1.1.1. Combustion Products

The presence of a number of chemical contaminants in the interior of a building is due to products of combustion. The use of stoves, heaters, dryers, refrigerators and oil burners facilitates the presence of oxides (CO, CO_2, NO, SO_2 and NO_2) in the air. Some of these contaminants can get into the air from outside sources due to inadequate air intakes. Among them stand out for their frequency as follows:

Table 3.1.1.1.1. Chemical contaminants in buildings

Combustion Products	Construction Materials	Consumer Products	Others
Carbon dioxide	Fibres	Particulate	Ozone
Carbon monoxide	VOCs	Pesticides	Metals
Smoke snuff			Radon

Carbon dioxide

Carbon dioxide is a gas that is formed by combustion of carbon-containing substances. With no local industry is the main source in the human respiratory and smoking. It is a suffocating simple presence whose absence leads to high concentrations of oxygen.

Carbon monoxide

Carbon monoxide is formed by incomplete combustion of carbon-containing substances. Their presence in media industry is not due to emission by internal combustion engines in garages within the building, taking inadequate fresh air outside and smoke. Has a stifling effect to join the haemoglobin in the blood and reduce the ability of oxygen to the tissues.

Smoke snuff

The fact is the smoke released into the air from a complex mixture of chemicals (known pollutants over 3000). Besides carbon monoxide, carbon dioxide and nitrogen oxides are produced like a wide variety of other gases and organic compounds.

3.1.1.1.2. Building Materials

Among the building materials used in the general isolation of the building, fibres, primarily of glass and asbestos, and different types of VOCs are of particular interest.

Introduction to Indoor Air Quality

Fibres

Fibreglass and asbestos are two types of fibres that present a potential contamination risk, whether generated in an industrial environment or in a non-industrial environment. Fibreglass is composed of amorphous glassy material. It is used as reinforcement in plastic, rubber, paper, fabric thermal insulation and air conditioning systems.

The term covers various forms of asbestos silicate minerals used in insulation materials normally. Although its use is prohibited or severely restricted in new buildings, it is still common in old buildings and can be a source of contamination during the execution of maintenance work and refurbishment, and as a result of the degradation of materials that contain them.

Volatile organic compound

Formaldehyde: Formaldehyde is widely used in the formulation of plastics, especially in the melamine-formaldehyde resin, urea-formaldehyde and phenol-formaldehyde used as insulation and varnishes. An inadequate design, poor curing and degradation produced over time, are the grounds for the issuance of this compound in ambient air. Formaldehyde can cause irritation and respiratory tract allergies and is considered a substance suspected to induce carcinogenic processes.

Solvents: Other materials which can be a source of pollution-generating chemicals in the air inside a building are furniture and wood decoration and rubber sealants agents, adhesives, coatings and textile materials.

3.1.1.1.3. Consumer Products Continually Coming through the User

They include products used in construction, such as paints, water-based (can contain mercury as a fungicide) and oil (oil), paints, plastics, adhesives, solvents, sealing products (many contain acetic anhydride) and coating fibre textiles, wallpaper and glue, as well as other new pesticides and repellents, cleaning products in general (including stain removers, oven cleaners and soaps for furniture and carpets), silicone polishes, cosmetics, deodorants, lacquers for polo and so on. Apart from the above mentioned organic compounds in building materials, consumer products can be categorized as particulates and pesticides.

Particulate

Particles that can be breathed could be respiratory irritant, especially for asthmatics. In industrial environments the main source of fine particles (2-3 mm) is cigarette smoke and aerosols from different types of sprays. The larger aerosol particles (3-1 mm) include leaking carpet fibres, skin flakes, dirt transported from outside, and so on. Often exposure to particles in the interior of a building is greater than outside.

Pesticides

This group includes a wide variety of dicumarinas, organophosphates, carbamates and chlorinated hydrocarbons that are used against insects, rodents and microbiological growth. While some are volatile and have a limited residence time, some can accumulate in dust and be redistributed. There are Unknown health effects associated with exposure to low concentrations of many pesticides.

3.1.1.1.4. Other Pollutants of Concern

Ozone

It is an oxidant that is present in certain conditions in outdoor air. In indoor air ozone is generated primarily from photocopying machines, high frequency discharge lamps, ultraviolet lamps and electric arc discharge.

Metals and metal compounds

The presence of lead is usually due to external sources. The presence of iron and manganese was also detected and its origin could not be justified. For its part, the air conditioning system releases Al_2O_3 powders from the corrosion of metal, which is built in the same way .

Radon

Some contaminants in the soil surrounding buildings can also infiltrate through the cracks in the foundations, such as radon. Radon is a radioactive gas from the decay of radium and belonging to the family of noble gases that emits alpha particles. Exposure to this issue has been linked to tissue damage and even cancer. Radon and its decay products are found in granite and phosphate deposits. In some cases may also be part of building materials.

3.1.1.2. Biological Contaminants

In the same way that chemical contaminants are considered microorganisms in the air inside should also be considered. To explain the production of biological aerosols we should refer to the concepts of reservoir, gearbox and disseminator. A reservoir is a medium that brings together a number of conditions that allow the organisms to survive in a certain environment, while the multiplier that promotes play and acts as a disseminator introducer of micro organisms and their metabolites in the air.

Biological contaminants, on the other hand, are basically classified as infectious agents, antigens and toxins because they form more usual.

3.1.1.2.1. Infectious Agents

Infectious diseases are transmitted more easily indoors than outdoors, because the volume of air in which micro organisms are diluted is lower, the contact is higher and people spend more time indoors than outside. We also need to consider that many diseases require direct contact between human hosts for transmission, while others, such as influenza, measles, smallpox, tuberculosis and some common colds are transmitted through the air, and the microorganism that make them up can easily survive their passage through the ventilation system if specific measures are not taken into account.

Other infectious diseases are transmitted directly from reservoirs into the environment. Among these are legionellosis pneumonia and other bacterial and most of the diseases caused by fungi. Legionella, for example, survive and multiply in cooling towers, humidifiers, shower heads, water and rubbish in general, which act as reservoirs for micro organisms and multipliers. Infectious diseases usually transmitted through the air can affect the respiratory system, at least initially, and the symptoms are manifested in both the upper and lower. Infectious agents can cause disease in any of those at risk, although the groups at highest risk are those with health problems and/or with a compromised immune system, especially children and the elderly. Sampling of infectious agents in air requires special equipment and experienced staff and is not very common. Much more common is the sample of infectious agents in the reservoirs and the multiplier.

3.1.2.2. Antigens

An Antigen is any substance which enters an animal organism with a mature immune system and is capable of provoking a specific immune response.

Most of the antigens that can be found in the air in enclosed environments are derived from micro organisms, arthropods or animals. Those present in the air can cause diseases such as hypersensitive pneumonia, allergic rhinitis and allergic asthma, among others.

The characteristic symptoms of pneumonitis hipersensitiva include fever, chills, choking, malaise and cough. The symptoms of allergic rhinitis are mucus, itchy eyes and nose and sinus congestion, while those of allergic asthma are shortness of breath and tightness in the chest as a result of constriction of the bronchi. Among the micro-reservoirs and multipliers for determining hypersensitivity diseases are substrates from outside, such as soil, plant material (live and not live) and water, and wet substrates from the own internal environment. Micro organisms can grow in any stagnant water and air. In the case of any dirty surface fungi can act as breeding, forming spores, which are directly exposed to the air and are dispersed throughout the building.

3.1.1.2.3. Toxins

Toxins are substances secreted by micro organisms that produce some adverse effects on living organisms. Most of the microbial toxins in the air of an atmosphere are made up of bacterial endotoxins and mycotoxins (from fungi). When the endotoxin-producing bacterium grows, releasing toxins into the water soluble (humidifier, for example), from which go to air. Endotoxins are associated with some symptoms of pneumonia. A distinctive smell of mold from the areas in which fungi are present is due to production from these volatile substances.

3.1.3. Legislation

Various international organizations such as WHO and CIBC (International Council of Building Research), closed the ASHRAE (American Society of Heating Refrigerating and Air Conditioning Engineers), and some countries like Sweden (The Swedish Council of Building Research), United States, Canada and Australia have developed guidelines and standards of exposure.

The air conditioning has to ensure that the air vent contains acceptably low concentrations of pollutants, which must be properly designed and maintained, since it can reduce pollutants to acceptable levels by dilution with clean air or eliminate foreign particles by filtration. According to ASHRAE, an acceptable indoor air is one in which there is no known contaminants in harmful concentrations as determined by the competent authorities and a substantial majority (80% or more) of staff not exposed to express dissatisfaction. Obviously, the definition is vague, not only with regard to acceptable levels, but also the concept of dissatisfaction.

There are no benchmarks to regulate the presence of microorganisms in the environment, The Committee for Bioaerosoles of ACGIH (American Conference of Governmental Industrial Hygienists) has recently published a Guide for the Assessment of bio aerosols in indoor environments that can be used as a starting point departure.

For those chemicals that do not have a reference value it is accepted (ASHRAE 62 [10]) that a concentration 1/10 TLV does not produce a significant increase in the number of complaints from members of a group of industrial work. Table 3.1.3.1 lists the maximum concentrations of pollutants that may be present in an outdoor air quality and represent a minimum so that it can be used for ventilation in an enclosed building. Table 3.1.3.2, is included for information and for a common air pollutant internal maximum exposure limits for OSHA and ACGIH (USA) in an industrial environment.

Table 3.1.3.1. Reference values of external air quality as EPA* = U.S. EPA Environmental Protection Agency

Contaminant	Long Exposure			Short Exposure		
	Mean concentration			Mean concentration		
	$\mu g/m^3$	ppm	Time	$\mu g/m^3$	ppm	Time
SO_2	80	0.03	1 year	365	0.14	24 hours
CO_2	-	-	-	40,000	35	1 hour
				10,000	9	8 hours
N_2	100	0.053	1 year	-	-	-
O_3	-	-	-	235	0.12	1 hour
Pb	1.5	-	3 months	-	-	-
Particulates	75	-	1 year	260	-	24 hours
Radon	0.2 Pico curies/l					

Table 3.1.3.2. Reference values and concentrations recommended for some industrial pollutants from (OSHA and ACGIH)

Contaminant	Concentration	Exposure Time	Origin
Asbestos	0.2-2.0 libres /cm^3	8 hours	TLV-TWA
SO$_2$	2 ppm	8 hours	PEL-TWA
	5 ppm	15 minutes	PEL-STEL
CO$_2$	10,000 ppm	8 hours	PEL-TWA
	5,000 ppm	8 hours	TLV-TWA
	30,000 ppm	15 minutes	PEL-STEL
NO$_2$	1 ppm	15 minutes	PEL-STEL
	3 ppm	8 hours	TLV-TWA
	5 ppm	15 minutes	PEL-STEL
Formaldehydes	1 ppm	8 hours	PEL-TWA
	2 ppm	15 minutes	TLV-STEL
CO	35 ppm	8 hours	PEL-TWA
	200 ppm	15 minutes	PEL-TECHO
	50 ppm	8 hours	TLV-TWA
	400 ppm	15 minutes	TLV-STEL
O$_3$	0.1 ppm	8 hours	PEL-TWA
	0.2 ppm	15 minutes	PEL-STEL
Pb	0.005 mg/m^3	8 hours	PEL-TWA
	0.15 mg/m^3	8 hours	TLV-TWA
PEL Permissible exposure limit TLV Threshold Limit Value			
TWA Time Weighted Average STEL Short Term Exposure Limit			

OSHA Occupational Safety and Health administration ACGIH American Conference of Governmental Industrial Hygienist.

3.1.2. Sensorial Load

Air quality can be perceived through the nasal sense and by its general chemical effects. These chemical effects appear as a general response of the nose and eyes mucous membranes to the irritant agents of the air. The combined response of these two senses determines whether the air is perceived as fresh and pleasant, or as flawed and irritating.

Taking as reference the sensory perception of air quality, it can be expressed in terms of the percentage of unsatisfied people who perceive the air as unpleasant when entering a confined space. Figure 3.1.2.1 shows the percentage of dissatisfied people with respects to the ventilation need of a standard person (office worker and sedentary adults with neutral thermal sensation). In this way, we can assess the air pollution caused by human biofluentes and express it in olf. The pollution intensity from most inside sources can be expressed with its equivalence to that emitted by a person. To do this we express the required olf needed to make the air very unpleasant and causing the same dissatisfaction as an actual pollution source. A decipol is the perceived quality of air in a confined space, with a contamination source of the intensity of an olf, but that is ventilated by 10 l/s of clean air, that is 1 decipol = 0.1 olf/(l/s). The curve in Figure 3.1 shows the percentage of dissatisfaction by the ventilation rate. This curve is based on studies conducted in Europe, where a group of 168 sedentary subjects (men and women) judged the quality of the polluted air. Similar studies of other research groups in the U.S. and Japan reached conclusions very close to these results. The Eq 3.1.2.1 gives the curve of Figures 3.1.2.1 and 3.1.2.2.

$$PD = 395 \cdot \exp(-1.83 \cdot q^{0.25}) \quad if \quad q \geq 0.32\,l/s \cdot olf$$
$$PD = 100 \cdot \exp(-1.83 \cdot q^{0.25}) \quad if \quad q < 0.32\,l/s \cdot olf$$

(3.1.2.1)

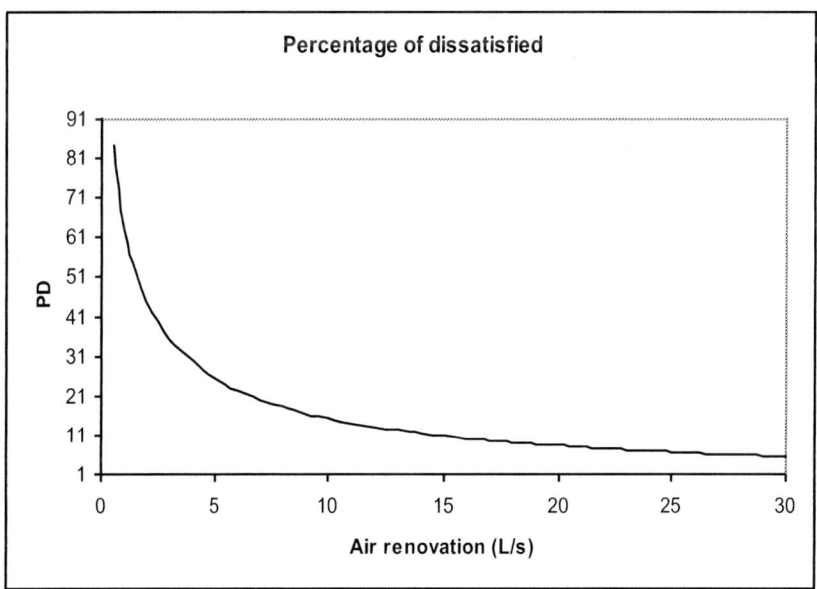

Figure 3.1.2.1. Percentage of dissatisfied for different ventilation rates.

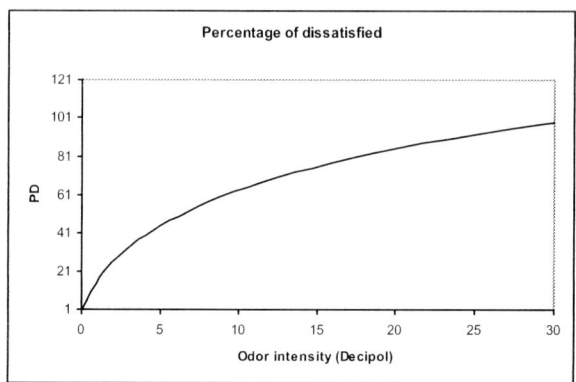

Figure 3.1.2.2. Relationship between perceived air quality and the percentage of dissatisfied.

Figure 3.1.2.2 shows the relationship between air quality expressed by the percentage of dissatisfied visitors and its corresponding decipol. It can also be expressed as the perceived air quality from dilution factors in relation to the defined smell levels. Furthermore, there were other measures of the pollution caused by other sources besides the human bioefluentes to express air quality perception. Table 3.1.2.1 shows the three indoor air quality levels according to the categories A, B and C. To determine the required ventilation from the comfort standpoint, it is essential to select the desired air quality level in the space to ventilate. In Table 3.1.2.1 three levels of perceived air quality have been proposed. Each level of quality corresponds to a certain percentage of dissatisfaction. In some areas, where a moderate demand exists, it may be sufficient to provide an air quality category of C, with 30% of dissatisfaction . The perceived air quality in Table 3.1.2.1 refers to the experiment of people when they initially entered an enclosed space as visitors. As the first impression is essential, it is very important that the immediately perceived air should be acceptable.

Table 3.1.2.1. Examples of the three levels of perception of air quality

Qualit level (category)	Quality of the perceived air		Req. Vent.*
	% Of unsatisfied	Decipol	l/s olf
A	10	0.6	16
B	20	1.4	7
C	30	2.5	4

*The ventilation examples are recounted exclusively to the quality of the perceived air.

Other indexes can be employed to evaluate the perception of indoor air quality and relate it with the percentage of locally unsatisfied, since it leads to an inadequate air cooling unacceptable for those who perceive it. From an energy point of view, this fact was related to the enthalpy, so the air will be more acceptable to low enthalpy. The origin of these concepts can be traced back to studies by Fanger. They conducted laboratory tests with 40 people exposed to air type conditions and assessed the air quality perceived if they had to work all day in this environment. This has come to the acceptability equation 3.1.2.2. The equation has been applied by Simonson [11] at temperatures of 20 °C and 24 °C.

$$A_{cc} = a \cdot h + b \qquad (3.1.2.2)$$

a and b are empirical coefficients whose values for clean air are a=-0033 and b = 1662.

From this equation we can deduce that indoor air acceptability increases at low enthalpy. When we want to condition indoor environments looking towards energy saving, we can reduce the ventilation rate but maintain a moderate enthalpy. That is, if the temperature raises 1°C, relative humidity has dropped by 5% to maintain the same acceptability. These processes have some limitations because when we are doing psychometric changes, we must bear in mind that over 50 kJ/kg air is unacceptable and that the relative humidity should be kept below 55% to reduce the proliferation of fungi [12].

Finally, the acceptability is also defined by the ASHRAE. It stipulates that this parameter is covered in the three central categories of the thermal sensation scale. In conclusion, the effect of humidity on the perception of indoor air quality is much higher than its effect on the thermal sensation due to the fact that it depends mainly on the enthalpy.

From these two indexes we can represent in a psychrometric chart the different lines of perception of indoor air quality and acceptability, as we can see in Figure 3.1.2.2.

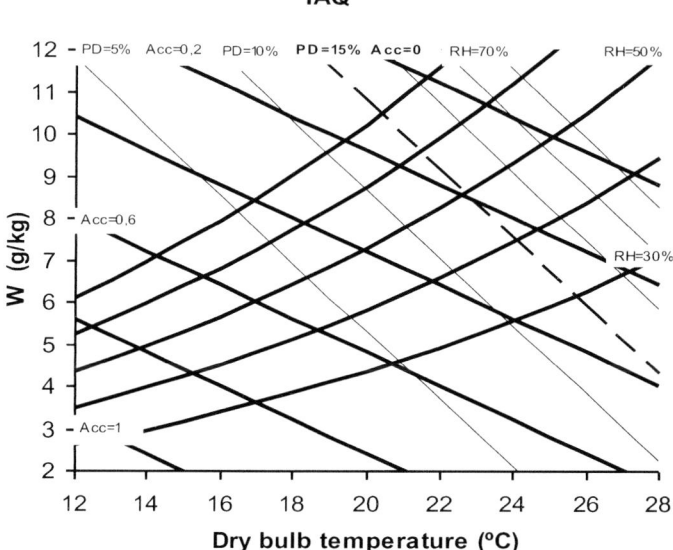

Figure 3.1.2.2. Perception of indoor air quality chart.

3.2. INDOOR AIR RENOVATION MODELS

The largest number of complaints about air quality inside a building within the area of thermal comfort and ventilation. According to the National Institute for Occupational Safety and Health (NIOSH), more than 50% of studies in buildings, the problems were caused by inadequate ventilation.

Thermal comfort is based on a balance between physical activity and clothing that is used on the one hand, relative humidity, temperature, air velocity and temperature radiant half on the other. The American Society of Heating, Refrigerating and Air Conditioning Engineers (ASHRAE) has developed standards for confined spaces, which should ensure the comfortability of 90% of the population. In general, the acceptable range of values is relatively narrow, given the relationship that exists between the two variables. A slight increase in air velocity, for example, can trigger a series of complaints while the temperature remains within acceptable limits.

Similarly, when the ventilation is incorrect as a result of an insufficient supply of fresh air from outside, there may be a source of accumulation of various pollutants to levels that are annoying to the occupants. The contribution of outside air should be sufficient to dilute the pollutants to levels that are below human perception, and obviously those considered harmful to health.

References since the middle of the eighteenth century recommended a minimum input of fresh air, per person, to dilute the concentration of human and bioefluentes avoid inconvenience due to bad smells.

In 70 years, ASHRAE published several papers recommending an injection of fresh air at least 34 m^3/h per person to prevent odours and an absolute minimum of 8.5 m^3/h per person to keep the concentration of carbon dioxide below 2500 ppm, which is half the average permissible exposure limit in a work environment. Most recently, the ASHRAE Standard 62-1989 recommends a minimum of 25.5 m^3/h per person for classrooms, 34 m^3/h for offices and 42.5 m^3/h for hospitals (diseased area). This standard also recommends an increase in the volume when there are problems in the mixture of air breathing zone or unusual sources of pollution. On the other hand, we must not forget that the primary purpose of an air conditioning system in an office building is to provide a good level of comfort.

Nowadays, there is great interest in determining the ventilation rate in indoor environments. In this sense, there have been various models to determine the air change to link it with the energy consumption [13].

Research techniques through the use of tracer gas to detect flaws in ventilation conditions are widespread. These tracer gas methods are grounded in mark the air in an environment with a gas that is identifiable to continue its movement. The tracer gas used in ventilation measures, is often colourless, odourless and inert, and should not be present in the normal atmosphere. This chapter will show the different models and sampling procedures for the renewal of indoor environments and their effect on local thermal comfort.

3.2.1. Techniques of Tracer Gas Monitoring

The techniques of tracer gases are the only ones that allow multiple types of quantitative assessments of ventilation, which include measures of air infiltration and renovation, the efficiency of extraction systems (foul air and smoke), as well as the dispersal of pollutants. It also can measure the air velocities of draughts in ventilation ducts.

Another advantage of tracer gas is the ability to perform actions in occupied environments. This is a very effective and accurate method since it takes into account the large effect of the occupation to assess the conditions for air renewal and the impact of opening and closing of doors and windows.

This will take into account the enormous impact of the occupation on the conditions for renewal of the air. The results of appropriate measures with tracer gas into a ventilation system will provide information on the amount of air introduced into each enclosure, the efficiency of heat recovery units, the amount of air taken from recycling, the "short circuit" effect of loaded air with the outside, and inlet distribution in the enclosures. Both the planning stage and the lack of regular checks can cause a great increase in energy consumption and, in many buildings, you can see the effect of SBS by not taking into account these factors.

The flow of air through a building or room can be assessed by one of the three tracer gas methods: concentration decay method, constant concentration method and constant emission method. The three methods are based on the continuity equation shown in Eq. 3.2.1.1 and in its possible simplifications.

$$V \frac{dC}{d\tau} = F(\tau) + N(\tau) \cdot C_{oa} - N(\tau) \cdot C(\tau) \qquad (3.2.1.1)$$

where;

V Volume of the air in the room (m^3).
C Tracer gas concentration in the air of the room (m^3/m^3).
τ Time (hours).
F(τ) Rate of introduction of tracer gas in the room (m^3/h).
C_{oa} Concentration of tracer gas in outdoor air (m^3/m^3).
N(τ) Air flow through the room (m^3/h).

This can be expressed as shows Eq. 3.2.1.2.

$$N(\tau) = \frac{F(\tau) - V \frac{dC}{d\tau}}{C(\tau) - C_{oa}} \qquad (3.2.1.2)$$

The number of renewals, N, are estimated dividing the flow rate through the compound by the volume of it, assuming that $C_{oa}=0$.

a) Concentration decay method

It is the most basic method used to measure air during renovations of discrete short periods. In this method a small amount of tracer gas is completely mixed with the air. To ensure uniformity of the tracer gas concentration at all points in the environment fans must be employed. After this, the concentration drop and the time needed to reach a concentration value close to zero must be measured.

When there are no new issues of tracer gas and the airflow rate of entry is constant, the concentration of tracer gas falls exponentially with time. Representing the natural logarithm of gas concentrations over time, we could get a line whose gradient is the linear speed of air the compound N, as is shown in Eq. 3.2.1.3. If we do not obtain a line we must consider that the tracer gas is not well mixed with the indoor environment and results are not valid.

$$N = \frac{\ln C(0) - \ln C(\tau_1)}{\tau_1} \qquad (3.2.1.3)$$

Finally, the equipment needed is a gas monitor, a tracer gas source and a fan to homogenise the mix.

b) Constant emission method

It is employed in long periods of continuous samples of renewal in simple areas. Tracer gas is emitted at a constant speed during the sampling period. Therefore, if the renovation and the concentration of tracer gas are constant, the number of renewals per hour, N will be in accordance with the Eq. 3.2.1.4.

$$N = \frac{F}{V \cdot C} \qquad (3.2.1.4)$$

The tracer gas concentration should be the same in all of the area at any given moment.

c) Constant concentration method

It is used to assess the ongoing renovations in occupied buildings. Using a gas monitor, tracer gas concentrations are measured in each zone. With this information the dosage of tracer gas to keep is controlled its concentration constant. A small fan can be used to facilitate the mixture.

Renewals are also expressed by the Eq. 3.2.1.4. The air renovation is proportional to the tracer gas emission required to maintain constant concentration. This approach has two advantages:

- Allows you to get the exact average speed of change in long periods and in situations where the air change varies over time.
- Can be used to assess changes in specific areas.

It is particularly appropriate to evaluate the continued outside air infiltration in each individual zone. The exchange of unwanted air between different parts of a building can also be assessed.

3.3. RELATED AIR QUALITY STANDARDS

As in previous chapters, it will be showed the actual standards related with IAQ.

a) ISO Standard
ISO 4225:1994 Air quality-General aspects-Vocabulary
ISO 4226:2007 Air quality-General aspects-Units of measurement
ISO 16813:2006 Building environment design-Indoor environment-General principles
ISO 16814:2008 Building environment design-Indoor air quality-Methods of expressing the quality of indoor air for human occupancy.
ISO 16000-8:2007 Indoor air-Part 8: Determination of local mean ages of air in buildings for characterizing ventilation conditions
ISO 16000-1:2004 Indoor air-Part 1: General aspects of sampling strategy.
ISO 16000-16:2008 Indoor air-Part 16: Detection and enumeration of moulds-Sampling by filtration.
ISO 16000-5:2007 Indoor air-Part 5: Sampling strategy for volatile organic compounds (VOCs).
ISO 7708:1995 Air quality-Particle size fraction definitions for health-related sampling.
ISO 8756:1994 Air quality-Handling of temperature, pressure and humidity data.

b) ASHRAE Standard
ANSI/ASHRAE 62.1-2004. Ventilation for Acceptable Indoor Air Quality.
ANSI/ASHRAE 62.2-2004. Ventilation and Acceptable Indoor Air Quality in Low-Rise Residential Buildings.

c) NTP [14]

NTP 343: New criterion for future indoors ventilation standards.
NTP 549: Carbon dioxide in evaluating indoor air quality.
NTP 345: Ventilation assessment using tracer gases.
NTP 243: Indoor air quality.
NTP 431: Characterization of indoor Air Quality.
NTP 488: Indoor air quality: Identification of Fungi.
NTP 299: Method for airborne bacteria and fungi counting.
NTP 335: Indoor air quality: pollen grains and fungi spores evaluation.
NTP 313: Indoor air quality: microbiological hazards in air conditioning and ventilation systems.
NTP 347: Chemical contamination: air concentration assessment.
NTP 315: Air quality: indoor low concentration gases.
NTP 521: Indoor air quality: Emissions from building materials and cleaning products.
NTP 289: Sick Building Syndrome: risk factors.
NTP 288: Building sick syndrome and building related diseases: bioareosol involvement.
NTP 290: The Sick Building Syndrome: questionnaire for its detection.
NTP 380: The Sick Building Syndrome: Simplified questionnaire.

REFERENCES

[1] World Health Organization. http://www.who.int/en/. Air quality guidelines for Europe Copenhague: WHO. 1987. Regional Office for Europe, European Series. 23.
[2] Guidelines for ventilation requirements in buildings Luxemburg: Office for publications of the European Communities, 1992.
[3] Weekes, D.M. and Gammage, R.B. (ed.) The practitioner's approach to indoor air quality investigations. 1990. *American industrial hygiene association*. Akron, Ohio.
[4] Cain, W.S. et al. Ventilation requirements in buildings-control of occupancy odor and tobacco smolke odor. 1983. Atmospheric environment. 17, 6, 1, 183-1.197.
[5] Berg-munch, B., Clausen, G. and Fanger, P.O. Ventilation requirements for the control of body odor in spaces occupied by women. 1986. *Environ. int.* 12, 195-199.

[6] Fanger, P.O. Introduction of the olf and the decipol units to quantify air pollution perceived by humans indoors and outdoors. 1988. *Energ. build.* 12, 1-7.
[7] Fanger, P.O., Lauridsen, J., Bluyssen, P. and Clausen, G. Air pollution sources in offices and assembly halls, quantified by the olf unit. 1988. *Energy build.* 12, 7-19.
[8] Fanger, P.O. The new comfort equation for indoor air quality. 1989. *ASHRAE Journal.* 10, 33-38.
[9] Cone, J.E. and Hodgson, M.J. (ed) Problem buildings: building-associated illness and the sick building syndrome. 1989. Occupational medicine. 4, 4.
[10] ANSI/ASHRAE 62.1-2004. Ventilation for Acceptable Indoor Air Quality. 2004.
[11] Simonson C. J., Salonvaara M. and Ojanen T. Improving Indoor Climate and Comfort with Wooden Structures. 2001. *Technical research centre of Finland.* Espoo.
[12] Viitanen, H. Factors affecting the development of mould and brown rot decay in wooden material and wooden structures. Effect of humidity, temperature and exposure time. 1996. Dissertation. Uppsala, SLU, For. Prod. 58.
[13] M.J. Cunningham. Using hygroscopic damping of relative humidity and vapour pressure fluctuations to measure room ventilation rates. 1994. Building and Environment, 29, 4, 501-510.
[14] Normativa Técnica de Prevención (NTP). Ministerio de Industria. http://www.insht.es/portal/site/Insht/menuitem.a82abc159115c8090128ca10060961ca/?vgnextoid=db2c46a815c83110VgnVCM100000dc0ca8c0RCRD

SECTION 2: RESEARCH TECHNIQUES

INTRODUCTION

Once the reader presents an adequate knowledge of the theory models of thermal comfort and indoor air quality and its evolution over the years, it is the moment to show the principal apparatus and its sampling methodology in accordance with the corresponding standards. Furthermore, in chapter 5 the principal software resources for the study of thermal comfort and indoor air quality conditions will be shown, among others such as energy saving and design corrections.

Chapter 4

MATERIALS AND METHODS

INTRODUCTION

This chapter describes the main sampling apparatus and methodology employed in the research of indoor environments to the development of some relevant research work. These devices can be classified into thermal comfort, ventilation and air monitoring equipment. Furthermore, fungi and mold were sampled and analysed in specific laboratories. Finally, information was added about portable dataloggers and climatic stations that are interesting to complement field studies.

4.1. THERMAL COMFORT

4.1.1. Thermal Comfort Datalogger

To calculate the thermal comfort indexes it is common to employ a thermal comfort data logger, as that shown in Figure 4.1. This data logger makes it easy to collect and analyze thermal comfort results once it connected to a PC or stored for later transfer to a computer. This dataloggers complies with ISO 7730; ISO 7726; ISO 7243; ASHRAE 55; SAE J2234 and ISO 14505. With this data the thermal comfort data logger will give us the Predicted Mean Vote (PMV), Percentage of Dissatisfaction (PPD), Draught Rating, Turbulence Intensity, Required Sweat Rate and the Operative, Equivalent, Effective and the Wet Bulb Globe Temperatures. To obtain these indexes, it is necessary to remember the need to sample 6 parameters to solve the thermal comfort equation derived by P.O. Fanger. All the parameters and measuring options were summarized in the Table 2.2.2.1 and its corresponding transducers will be shown in the next section.

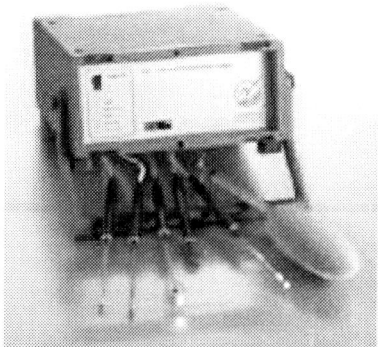

Figure 4.1.1.1. Thermal comfort module (Courtesy of LumaSense Technologies[1]).

4.1.2. Air Temperature Transducer

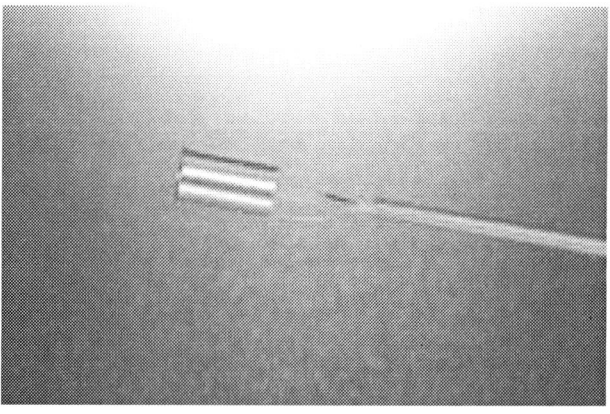

Figure 4.1.2.1. Air Temperature Transducer INNOVA MM0034 (Courtesy of LumaSense Technologies).

The air temperature transducer of Figure 4.1.2.1.1 measures the air temperature with minimal thermal radiation interference from hot or cold objects. It measurement principle provides accurate measurement results, which are both stable and traceable.

4.1.3. Surface Temperature Transducer

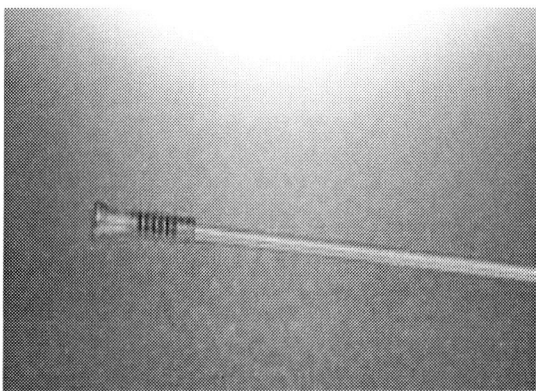

Figure 4.3.1. Surface Temperature Transducer - INNOVA MM0035 (Courtesy of LumaSense Technologies).

The surface temperature transducer in Figure 4.3.1 measures the actual temperature of the given object. In order to do this, it is essential to keep the contact surface area of the transducer as small as possible. This prevents the transducer from affecting the temperature of the object while, at the same time, being able to maintain a good contact between the sensor and the object.

The results provided by this transducer are used to evaluate thermal discomfort from floors as well as providing input for mean radiant and plane radiant temperature calculations.

4.1.4. Radiant Temperature Asymmetry Transducer

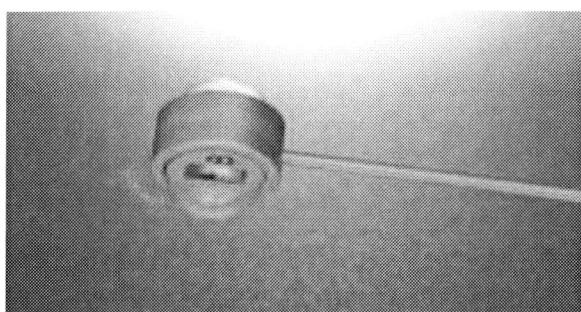

Figure 4.1.4.1. Radiant Temperature Asymmetry Transducer INNOVA MM0036 (Courtesy of LumaSense Technologies).

The radiant temperature asymmetry transducer in Figure 4.1.4 consists of two identical surfaces that independently measure net incident radiation on each plane surface. Measurement results from the transducer are used to evaluate radiant asymmetry discomfort from hot and cold surfaces (according to ISO 7730 [2]).

4.1.5. Humidity Transducer

Figure 4.1.5.1. Humidity Transducer - INNOVA MM0037 (Courtesy of LumaSense Technologies).

The humidity transducer in Figure 4.1.5.1 measures the absolute humidity of air. The design provides a robust and very stable transducer for field use that comprises a light emitting diode (LED), a light sensitive transducer, a mirror, a cooling element and a Pt100 temperature. A cooling element is attached to the conical mirror. When the humidity measurements are started, the cooling element is activated and the temperature of the mirror begins to drop. Control of the transducer now moves to the LED and light-sensitive transistor. The LED emits a constant beam of light, which is reflected by the conical mirror. Under normal circumstances, the light-sensitive transistor does not receive any of the light being emitted by the LED, however, as the temperature of the mirror drops condensation forms on the surface. The mirror is reflecting the light which now becomes scattered and is detected by the light-sensitive transistor. This transistor now controls the temperature of the mirror so that there is a constant film of dew on the mirror's surface. The Pt100 sensor registers this dew-point temperature.

4.1.6.1. Air Velocity Transducer

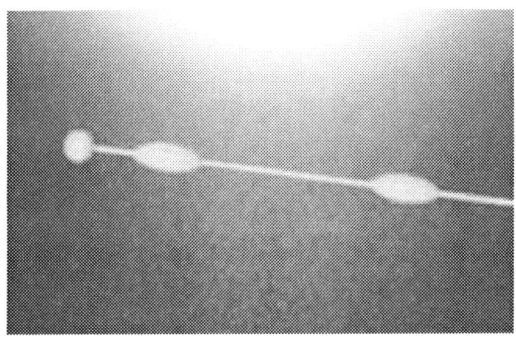

Figure 4.1.6.1. Air Velocity Transducer-INNOVA MM0038 (Courtesy of LumaSense Technologies).

The air velocity transducer in Figure 4.1.6.1 is based on the constant temperature difference anemometer principle and is based on the fact that air velocity is measured as a function of heat loss from a heated body, by measuring the power input required to maintain a constant temperature between two sensor elements. Heat loss is, however, also a function of the temperature and direction of airflow and the radioactive exchange with the surroundings.

4.1.7. Operative Temperature Transducer

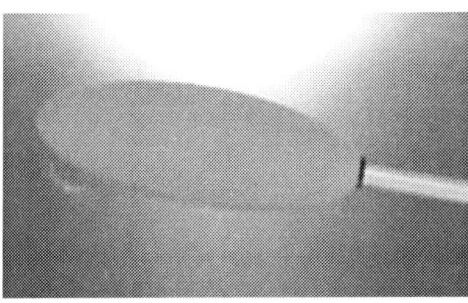

Figure 4.1.7.1. Operative Temperature Transducer - INNOVA MM0060 (Courtesy of LumaSense Technologies).

Normally, the amount of heat given off by a human body through radiation is approximately the same as the amount of heat given off by convection. Therefore, a simple air temperature measurement is a bad indication of the thermal environment. Operative temperature takes both radiation and convection into account and is therefore a much better indicator. To sample the operative temperature we must consider that people do not maintain the same posture. For this reason, the transducers present three settings: vertical, 30° from the vertical, and horizontal, which represent the body in the standing, sitting and lying positions respectively as we can see in Figure 4.1.7.1.

4.1.8. Equivalent Temperature/Dry Heat Loss Transducer

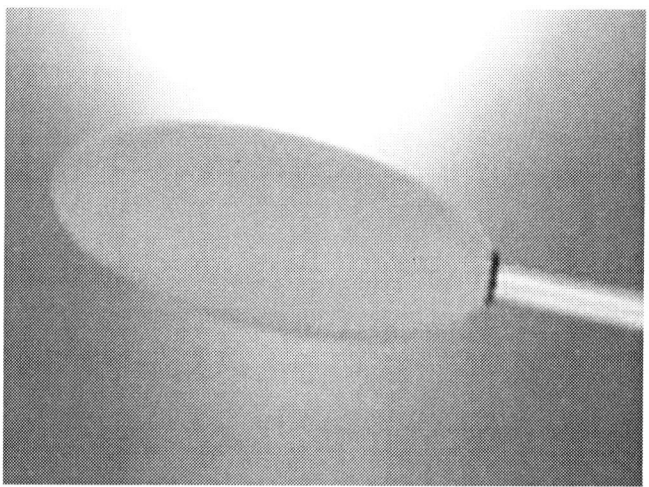

Figure 4.1.8.1. Dry Heat Loss Transducer - INNOVA MM0057 (Courtesy of LumaSense Technologies).

The Dry Heat Loss Transducer in Figure 4.1.8.1 enables you to simulate the human body thermally while it measures the Equivalent Temperature. In consequence, with this transducer, the integrated effect on the body of the air temperature, the mean radiant temperature and the air velocity can be evaluated. The human body's senses are only able to feel heat loss from the body. The dry heat loss transducer is able to measure part of this basic parameter directly; the other part of the heat loss is the evaporative heat loss. This can be estimated by measuring the humidity of the air. By measuring these two parameters it is enough for a thermal comfort evaluation as was shown in Table 2.2.2.1.

4.2. HEAT STRESS

4.2.1. Wet Bulb Globe Temperature – WBGT

Figure 4.2.1.1. Wet Bulb Globe Temperature – WBGT (Courtesy of LumaSense Technologies).

As was shown in 2.4, the measurement of three transducers are required for the analytical determination and interpretation of heat and cooling stress in accordance with ISO 11079:2007 and ISO 7933:2004. These transducers are represented in Figure 4.2.1.1.

Natural Wet Bulb Temperature
Perspiration is one of the ways in which the body controls its internal temperature. As the perspiration evaporates it has a cooling effect on the body. The natural wet bulb temperature simulates the evaporative heat loss effect. The sensor achieves this by having an unshielded bulb covered with a wet cotton sock or wick. A 40 ml reservoir of distilled water ensures that the wick is kept moist. The evaporation from the wick cools the sensor in the same way that sweat cools the body. Because of this cooling effect the natural wet bulb temperature is normally lower than the air temperature. Note that the natural wet bulb temperature is not the same as the psychrometric wet bulb temperature. The dimensions of the sensor are defined by ISO 7243.

Globe Temperature

This temperature indicates the amount of heat exchanged by the body due to radiation. The transducer consists of a Pt100 temperature-sensing element situated at the centre of the globe (150 mm in diameter), which is made of 0.4 mm copper sheet coated with optically black lacquer that has an emission coefficient of 0.98. The temperature measured at the centre is an equilibrium temperature caused by radioactive and convective heat exchanges between the globe and the environment. As a result of this, the globe temperature is influenced by air velocity, air temperature and radiant temperature. Size and construction of the globe are defined by ISO 7243.

Air Temperature

Because the globe over-estimates the influence of direct sunshine, it is necessary to provide the true air temperature as well. This is measured by a platinum (Pt100) sensor, which is radiantly shielded. The sensor is surround by an open-ended aluminium-foil cylinder. This is highly polished to reduce the thermal radiation interference from any hot or cold bodies in close proximity of the transducer. The cylinder with its open ends enables a free flow of air to come in contact with the sensor.

4.3. INDOOR AIR QUALITY AND AIR RENOVATION

4.3.1. Gas Monitoring

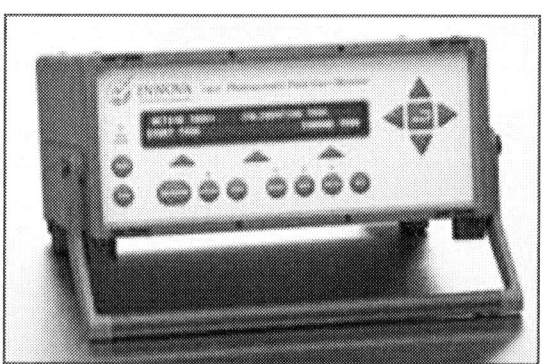

Figure 4.3.1.1. Photoacoustic Field Gas-Monitor Innova -1412 (Courtesy of LumaSense Technologies).

Materials and Methods 75

The field gas-monitor in Figure 4.3.1.1 is a highly accurate, reliable and stable quantitative gas monitor. Its measurement principle is based on the photo acoustic infrared detection method. This means that it can measure almost any gas that absorbs infrared light. Appropriate optical filters are installed in a filter carousel so that it can selectively measure the concentration of up to 5 component gases and water vapour in any air sample. Further, the instrument can compensate for interference between the measured gases, using the cross-compensation feature. In consequence, this gas monitor lets us sample indoor air-quality and ventilation using a tracer-gas in accordance with 3.2 and 3.1.3 indications.

4.4. PORTABLE DATALOGGERS

4.4.1. Thermal Comfort Dataloggers

Some times it is necessary to take field samples for long periods of time. In these cases it is necessary to employ portable dataloggers for temperature, relative humidity, and air velocity like those shown in Figures 4.4.1, 4.4.2 and 4.4.3. These dataloggers need less space than modules shown before and, in consequence, could be located near the worker during longs periods of time.

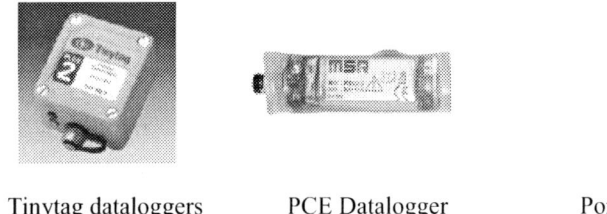

Tinytag dataloggers   PCE Datalogger   Portable Air velocity
(Courtesy TinyTag)   (Courtesy of PCE Ibérica) transducer PCE-009
                                                (Courtesy of PCE Ibérica)

Figure 4.4.1.1. Portable temperature, relative humidity and air velocity dataloggers.

4.4.2. Work Risk Prevention Dataloggers

In working places areas is usual to found extreme conditions that are upper ISO 7933:2004 indications. In consequence, portable heat stress and sound level meters like that showed in Figure 4.4.2.1 must be employed to get a field control of indoor ambiences.

Portable Heat stress transducer
(Courtesy of PCE Ibérica)

Digital Sound Level Meter
(Courtesy of Casella Ltd)

Figure 4.4.2.1. Portable work risk prevention dataloggers.

4.4.3. IAQ datalogger

A field study of indoor ambiences requires, in some situations, a portable sample of ambience conditions. In particular CO_2 and dust data loggers will let us define the air renovation and its air quality as is showed in Figure 4.4.3.1. Furthermore, it is known that fungi growth happens especially in particular zones of a building.

4.4.4. Other Dataloggers

4.4.4.1. Weather Stations

Once indoor conditions have been sampled, they must be connected with outdoor conditions for different applications, such as energy saving in buildings. These outdoor conditions may be obtained by nearby weather stations that sample the principal outdoor parameters like temperature and relative humidity, as we can see in Figure 4.7.1.

Materials and Methods

  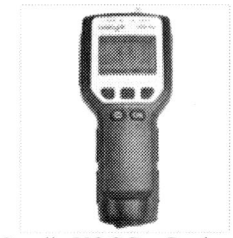

Portable CO$_2$ IAQ910 transducer (Courtesy of PCE Ibérica) Casella Microdust (Courtesy of Casella Ltd) Casella VOC Pro Real (Courtesy of Casella Ltd

Figure 4.4.3.1. Portable IAQ dataloggers.

Figure 4.4.4.1.1. Casella Fixed Automatic Weather Stations (Courtesy of Casella Ltd).

4.5. MEASUREMENT POSITIONS

4.5.1. Thermal Comfort Measurements Location.

In accordance with ASHRAE Standard [10], measurements shall be made in occupied zones of the building at locations where the occupants are known to or are expected to spend their time and, in unoccupied rooms. The evaluator should make a good faith estimate of the most significant future occupant locations within the room and make appropriate measurements. If occupancy distribution cannot be estimated, then the measurement locations shall be in the center of the room or zone and 1.0 m inward from the center of each of the room's walls.

In the case of exterior walls with windows, the measurement location should be 1.0 m inward from the center of the largest window. In either case, measurements should be taken in locations where the most extreme values of the thermal parameters are estimated or observed to occur but allowing a proper air circulation around measurement sensors with positions as described below.

Absolute humidity need only be determined at one location within the occupied zone in each occupied room as there is no reason to expect large humidity variations within that space.

4.5.2 Thermal Comfort Measurements Height Above Floor

Air temperature and air speed shall be measured at e 0.1, 0.6 and 1.1 m levels for sedentary occupants at the locations specified before. Standing activity measurements shall be made at 0.1, 1.1 and 1.7 m levels. Operative temperature or PMV-PPD shall be measured or calculated at 0.6 m level for seated and 1.1 m level for standing occupants.

Radiant asymmetry shall be measured at 0.6 m level for seated and 1.1 m for standing occupants. If desk-level furniture blocks the view of strong radiant sources and sinks, the measurements are to be taken above desktop level. Floor surface temperatures must be measured with the anticipated floor coverings installed and humidity shall be measured at any level within the occupied zone.

4.5.3. Ventilation and Air-Exchange Measurements

In this case, the doser system marks the supply-air supply of the room with a known amount of tracer-gas such as Sulphur Hexa Fluoride (SF_6), Freon 134a, Freon 152 or Nitrous Oxide (N_2O). A measurement of Indoor Air Quality (IAQ) is an important application for detection of Carbon Dioxide (CO_2), Carbon Monoxide (CO) and Volatile Organic Compounds (VOC).

The sampler system then takes a sample of the return-air from the room, and delivers the sample to the IAQ module for analysis. While the module performs one analysis, the multi sampler takes the next sample for analysis from the room. As the amount of tracer-gas delivered to the room is known, and the remaining concentration of tracer-gas in the samples is determined, the ventilation-system performance can be automatically calculated as was shown in 3.2.

4.6. RELATED SAMPLING STANDARDS

Now it will be numbered the principal actual standards about sampling methodology:

ISO 14956:2002 Air quality-Evaluation of the suitability of a measurement procedure by comparison with a required measurement uncertainty
ISO 11222:2002 Air quality-Determination of the uncertainty of the time average of air quality measurements
ISO 13752:1998 Air quality-Assessment of uncertainty of a measurement method under field conditions using a second method as reference
ISO 9359:1989 Air quality-Stratified sampling method for assessment of ambient air quality.
ISO 9169:2006 Air quality-Definition and determination of performance characteristics of an automatic measuring system

REFERENCES

[1] Lumasense Technologies. http://www.lumasense.dk/ [Accessed December 2010]
[2] ISO Standard 7730: Moderate Thermal Environments-Determination of the PMV and PPD indices and specification of the conditions for thermal comfort.
[3] ISO Standard 7933: Ergonomics of the thermal environment-Analytical determination and interpretation of heat stress using calculation of the predicted heat strain.
[4] ISO Standard 7726: Ergonomics of the thermal environment-Instruments for measuring physical quantities.

[5] ISO Standard 7243: Hot environments-Estimation of the heat stress on working man, based on the WBGT-index (wet bulb globe temperature).
[6] ISO Standard 14505: Evaluation of thermal environment in vehicles. Part 1; Principles and methods for assessment of thermal stress. Part2; Determination of Equivalent Temperature.
[7] Tinytag. http://www.geminidataloggers.com/contact-us/thank-you.
[8] PCE Group. http://www.pce-iberica.es. [Accessed December 2010]
[9] Casella http://casella-es.com/principal/principal.htm. [Accessed December 2010]
[10] ASHRAE Standard 55-2004-Thermal Environmental Conditions for Human Occupancy. 2004.

Chapter 5

SOFTWARE RESOURCES

INTRODUCTION

An easier way to solve building environment equations is to use a building energy simulation tool. These computer programs can simulate a building and its HVAC system. They can predict results such as the heating and cooling loads of a building, as well as the indoor thermal climate of a space [1]. Before the software package can accurately simulate a building, there are some key aspects that need be entered into the program. These include the geometry of the building and the number of zones that make up the building, as well as the internal loads within the building.

Another way in which building energy tools vary is in the type of simulations that they can perform. Most programs can calculate peak heating and cooling loads, the total energy consumption, system performance and costs. Many programs also estimate the indoor temperature. However, the capability to predict the indoor humidity is also required by the scientific community like IEA in his Annex 41.

Using the Building Energy Software Tools Directory [2], (a website listing 265 different energy related software tools), a list was made of all software packages that might have the desired capabilities. From that software a list of software related with thermal comfort and another list related with Indoor Air Quality were drawn up and are shown in the last section of this chapter. In the next sections the principal software resources will be analysed. Special interest will be shown to HAM-tools and, in consequence, it will have its own section in this chapter.

5.1. ACTUAL BUILDING SIMULATION SOFTWARE

5.1.1. Blast

The Building Loads Analysis and System Thermodynamics (BLAST) [3] provide a simulation of energy consumption, system performance and cost. As well, it also calculates hourly heating and cooling loads based on a fundamental heat balance method. Information on the building structure can be input from a library, which has data from ASHRAE Fundamentals. The disadvantage that this software presents is that it requires a high expertise level and that it may not assimilate humidity transfer through the covering of the buildings.

5.1.2. BSim

The BSim family of computer simulation tools was created by Danish Building and Urban Research [4] and a database with moisture properties of different materials can be added to into the program. This simulates the moisture within the building construction. This building simulation tool has the ability to analyze complex buildings, as well as those with special requirements for the indoor climate.

5.1.3. CHVAC

CHVAC is a product of Elite Software [5]. It is a simulation tool for calculating peak heating and cooling loads for commercial buildings. The building loads are calculated using the somewhat outdated CLTD procedures as described in the 1997 ASHRAE Handbook of Fundamentals. This allows the program to perform calculations for buildings of any orientation and that the building to be analyzed can be broken down into different zones. Its disadvantages are that it can only calculate design loads and not perform an energy analysis on the building and cannot simulate moisture storage.

5.1.4. DOE-2

The DOE-2 [6] building simulation package was developed by the Department of Energy. The program simulates the energy performance of the building, including the life cycle cost of operating the HVAC system. Has been extensively validated by comparison with actual measurements and calculations and, in consequence, comes highly recommended by ASHRAE as a complete energy simulation tool. The equations used to perform all calculations within the program are based on ASHRAE published algorithms.

Its principal disadvantage is that the input into the program presents a high knowledge requirement from the user, and it does not have the ability to simulate IAQ and moisture storage.

5.1.5. Energyplus

The Building Systems Laboratory together with Lawrence Berkeley National Laboratory and the Department of Energy has combined two programs, BLAST and DOE-2 [7]. With this program, the heating, cooling, lighting, ventilating and other energy related flows in a building can be simulated. It uses a heat balance-based zone simulation method to perform calculations. When analyzing buildings, EnergyPlus can account for moisture adsorption and desorption within the building elements. A major disadvantage of the program is that it was not designed to be the main graphical interface used making the program more difficult to use.

5.1.6. HAP

Hourly Analysis Program (HAP) from Carrier [8] is a load estimating simulation tool. It provides results of both building loads and equipment operation for commercial buildings. All load calculations are performed using the Transfer Function load method.

Its principal disadvantage is that it does not have access to the source code making it somewhat inadequate for researchers, and cannot analyze moisture flow through constructions.

5.1.7. TRNsys

The TRaNsient SYstems Simulation (TRNSYS) [9] program was developed by the Solar Energy Laboratory at the University of Wisconsin, in 1975. It is a flexible simulation tool that can simulate the transient performance of thermal energy systems. The simulation program does not make any assumptions about the building or system being used so the user must have all the necessary details required for input into the program. Has the ability to calculate indoor temperature and relative humidity and permits the user to develop and run models of other building components and systems.

5.1.8. UMIDUS

UMIDUS [10] was developed at the Thermal Systems Laboratory of Pontifical Catholic University of Parana in Brazil. This program has the capability to analyze the hygrothermal performance of building elements taking into account one-dimensional heat and moisture transfer. Furthermore, it has the ability to predict moisture and temperature profiles within different layers of the construction but is not able to predict relative indoor humidity. This program is free to download from the Internet.

5.2. HAM-TOOLS

5.2.1. Advantages of HAM-tools

As was shown, a host of commercially available computer tool models already exist for modelling single components or whole buildings. For modelling whole buildings, there are models for the hourly energy balance like Bsim1, ESP-r2, EnergyPlus3, and more. While these tools are fully appropriate for designing standard buildings, they are not suitable for modelling innovative building elements such as building integrated heating and cooling systems, ventilated glass facades and solar walls, as these have not been defined in the program [11].

A different approach is the use of modular simulation tools, represented i.e. by TRNSYS4 and SPARK5. In Strand et al. (2002), the differences in the two types of tools are discussed. Here it was found that the major shortcomings of building energy simulation programs have so far been the inability to accurately model HVAC systems that are not "standard". This argumentation can easily be expanded to include advanced building elements. Modular models, on the other hand, have the advantage that components and systems can be modelled as the need appears. In addition, transparency of the existing components is essential, if the user/developer wishes to implement any modifications. A transparent, modular and open source system for modelling heat and moisture flows in buildings should therefore be a user-friendly tool that can be expanded as needed in the future.

The above-mentioned concerns have given rise to the need to develop an open and freely available building physics toolbox among Authors. The beginning of the International Building Physics Toobox (IBPT) was laid down by two groups of researchers working independently of each other developing building physics models in Simulink. For both groups, the reason for starting to use Simulink as the development environment was a need to model, in great detail, the processes of heat, air and moisture transfer. In both groups the reason for choosing Simulink, which is part of the Matlab package, was a large degree of flexibility, modular structure, transparency of the models and ease of use in the modelling process.

Simulink has already previously been used by other research communities (SIMBAD and CARNOT), but the models have either not been an open source, free of cost or have not been directly applicable to building physics modelling.

The modular structure in Simulink - using systems and subsystems - makes it easier to maintain an overview of the models, and new models can just as easily be added to the pool of existing models. Another advantage of using Simulink is the graphical programming language based on blocks with different properties such as arithmetic functions, input/output, data handling, transfer functions, state space models and more. Furthermore, Simulink has built-in state of the art ordinary differential equation (ODE) solvers, which are automatically configured at run-time of the model. Therefore, only the physical model needs to be implemented, and not the solver. Further, models can be created using a number of different approaches. These include assembling models directly in Simulink using the standard blocks, Matlab m-files, S-functions, and Femlab9 models using one-, two-, or three-dimensional finite element calculations. This wide variety of modeling techniques with different advantages and disadvantages means, that the optimal choice can always be made with respect to the task.

The graphical approach also makes it easy to show the very complex interaction between the different parts of the model. In addition, people who are not used to programming can easily get started building their own models or altering existing ones. Therefore, the toolbox also represents a good way to teach building physics.

Comparing IBPT to other modular building energy simulation programs (i.e. TRNSYS, SPARK and EnergyPlus), a few notes can be made. (1) IBPT does not require the same level of detailed programming knowledge (but advanced programming is possible), (2) it can use Simulink's built in solvers (but new solvers can be added by the user), (3) a large degree of integration using the Matlab package can be achieved and as described below (4) all programming codes are freely available from the website [11].

5.3. HAM TOOLS SIMULATION

The mathematical model employed in this simulation is the result of whole building Heat, Air and Moisture (HAM) [12, 13, 14] balance and depends on moisture generated from occupant activities, moisture input or removed by ventilation, and moisture transported and exchanged between indoor air and the envelope [15].

The mathematical model is based in the numerical resolution of the energy and moisture balance through the building. In accordance with the next equations [13], the heat flow presents a conductive and a convective part as we can see in the Eq. 5.3.1 and described in Eq. 5.3.2 and 5.3.3.

$$q = q_{conductive} + q_{convective} \quad (5.3.1)$$

$$q_{conductive} = -\lambda \frac{\partial T}{\partial x} \quad (5.3.2)$$

$$q_{convective} = m_a \cdot c_{pa} \cdot T + h_{evap} \quad (5.3.3)$$

where:
  λ is the thermal conductivity (W/mK)
  T is the temperature (°C)
  m_a is the density of moisture flow rate of dry air (kg/m^{2}s)
  c_{pa} is the specific heat capacity of the dry air (J/kg K)

h_{evap} is the latent heat of evaporation (J/kg)

The moisture flow transfer was separated into liquid and vapour phases as we can see in Eq. 5.3.4 and 5.3.5.

$$m_l = K \cdot \frac{\partial P_{suc}}{\partial x} \qquad (5.3.4)$$

where;

m_l is the density of moisture flow rate of vapour phase (kg/m²s)

K is the hydraulic conductivity

P_{suc} is the suction pressure (Pa)

The vapour phase was divided into diffusion and convection as we can see in Eq. 5.3.5.

$$m_v = -\delta_p \cdot \frac{\partial p}{\partial x} + m_a \cdot x_a \qquad (5.3.5)$$

where;

δ_p is the moisture permeability (s)

x_a is the water vapour content (kg/kg)

The mass airflow through the structure driven by air pressure differences across the structure is shown in Eq. 5.3.6.

$$m_a = r_a \cdot \rho_a \qquad (5.3.6)$$

where;

r_a is the density of the air flow rate (m³/m²s)

ρ_a is the density of the material (kg/m³)

The final energy and moisture balances are shown in Eq. 5.3.7 and 5.3.8.

$$-\frac{\partial}{\partial x} q = c \cdot \rho_o \cdot \frac{\partial T}{\partial t} \qquad (5.3.7)$$

$$-\frac{\delta}{\partial x} m = \frac{\partial w}{\partial t} \qquad (5.3.8)$$

where

ρ_o is the density of the dry material (kg/m³)
c is the specific heat capacity of the material (J/kg K)
w is the moisture content mass by volume (kg/m³)
t time (s)
x space coordinates (m)

Now, the numerical model, based on a control volume method present lumped the thermal capacity C in the middle of the total thickness d/2 and, in consequence, the thermal resistances for one half are shown in equations 5.3.9, 5.3.10 and 5.3.11.

$$R = \frac{d/2}{\lambda} \qquad (5.3.9)$$

$$R_p = \frac{d/2}{\delta_p} \qquad (5.3.10)$$

$$R_{suc} = \frac{d/2}{K_{suc}} \qquad (5.3.11)$$

The obtained discretized heat and moisture balance equations are shown in equations 5.3.12 and 5.3.13.

$$\frac{T_i^{n+1} - T_i^n}{\Delta t} = \frac{1}{C^n} \cdot \left\{ \left[\frac{(T_{i-1} - T_i)}{R_{i-1} + R_i} + \frac{(T_{i+1} - T_i)}{R_{i+1} + R_i} \right] - h_{evap} \cdot \left[\frac{(p_{i-1} - p_i)}{R_{p,i-1} + R_{p,i}} + \frac{(p_{i+1} - p_i)}{R_{p,i+1} + R_{p,i}} \right] \right\} \ldots$$

$$+ \begin{cases} m_a \cdot c_{pa} \cdot (T_{i-1} - T_i)^n, m_a > 0 \\ m_a \cdot c_{pa} \cdot (T_i - T_{i+1})^n, m_a < 0 \end{cases}$$

$$(5.3.12)$$

$$\frac{w_i^{n+1} - w_i^n}{\Delta t} = \frac{1}{d} \cdot \left\{ \left[\frac{(p_{i-1} - p_i)}{R_{p,i-1} + R_{p,i}} + \frac{(p_{i+1} - p_i)}{R_{p,i+1} + R_{p,i}} \right] - \left[\frac{(P_{suc,i-1} - P_{isuc,i})}{R_{suc,i-1} + R_{suc,i}} + \frac{(P_{suc,i+1} - p_{suc,i})}{R_{suc,i+1} + R_{suc,i}} \right] \right\} \ldots$$

$$+ \begin{cases} 6.21 \cdot 10^{-6} \cdot m_a \cdot (p_{i-1} - p_i)^n, m_a > 0 \\ 6.21 \cdot 10^{-6} \cdot m_a \cdot (p_i - p_{i+1})^n, m_a < 0 \end{cases}$$

(5.3.13)

where i is the objective node and i+1 and i-1 are the preceding and following node and n and n+1 the previous and corresponding time steps.

To solve these balance equations room models were created from the individual Building Physics Toolbox [15, 16]. Ham–tools library is a Simulink model upgraded version of H-Tools with a similar structure and specially constructed for thermal system analysis in building physics. The library contains blocks for 1-D calculation of Heat, Air and Moisture transfer thought the building envelope components and ventilated spaces. The library is the part of IBPT-International Building Physics Toolbox, and available for downloading for free [17].

This library presents two main blocks; a building envelope construction (walls, windows) and thermal zone (ventilated spaces), which are enclosed by the building envelope. Component models provide detailed calculations of the hydrothermal state of each subcomponent in the structure; according to the surrounding conditions to which it is exposed.

In Figure 5.3.1 we can see the principal blocks employed for a building simulation. There we can see a block that represents the different exterior/interior walls, floor, roof and windows components. These constructions are defined respecting their physical properties (density of the dry material and open porosity), thermal properties (specific heat capacity of the dry material and thermal conductivity) and moisture properties (sorption isotherm, moisture capacity, water vapour permeability and liquid water conductivity) in accordance with the BESTEST structure. Other parameters are considered in the heat and moisture building balance, for example, internal gains (convective gains, radioactive gains and moisture gains), air change and heating/ cooling system.

Building's characteristics are defined in the thermal zone block indicating the surface areas, orientations and tilts of each wall. Room volume, ambient air gain from the heat originated from solar energy and initial temperature is adjusted so. The Thermal model of the classroom is based on the WAVO model described by de Witt (2000) [2] and developed by the assumptions that long wave radiation is equally distributed over the walls; the room air has a uniform temperature, the surface coefficients for convection and radiation are constant and, finally, that all radioactive heat input is distributed in such a way that all the surfaces, except windows, absorb the same amount of that energy per unit of surface area.

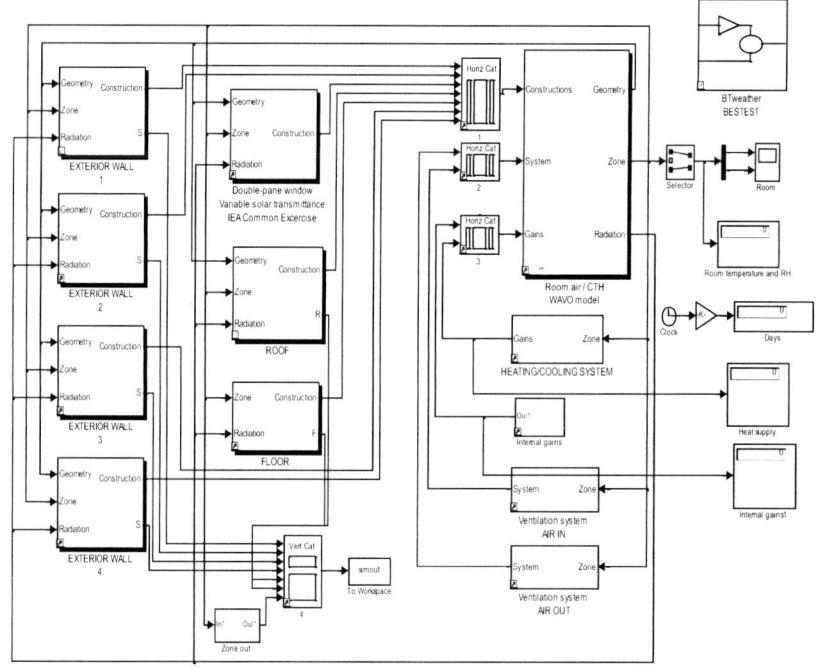

Figure 5.3.1. Matlab blocks for buildings simulations.

5.4. SOFTWARE

In the next section it will be showed the actual software resources to study the thermal comfort and indoor air quality in buildings in accordance with the U.S. Department of Energy [18].

Table 5.4.1.1. Thermal comfort and energy saving software

Tool	Applications	Cost
1D-HAM	Heat, air, moisture transport, walls	
AFT Mercury	Optimization, pipe optimization, pump selection, duct design, duct sizing, chilled water systems, hot water systems	
AkWarm	Home energy rating systems, home energy, residential modelling, weatherization	
Apache	Thermal design, thermal analysis, energy simulation, dynamic simulation, system simulation	
ApacheHVAC	Buildings, HVAC, simulation, energy performance	
ApacheSim	Thermal simulation, energy consumption	
AUDIT	Operating cost, bin data, residential, commercial	
BEAVER	Energy simulation, thermal analysis	
BSim2002	Building simulation, energy, daylight, thermal analysis, indoor climate	
Building Design Advisor	Design, daylighting, energy performance, prototypes, case studies, commercial buildings	Free
Building Energy Analyzer	Air-conditioning, heating, on-site power generation, heat recovery, CHP, BCHP.	
Building Energy Modelling and Simulation: Self-Learning Modules	Energy simulation, buildings, courseware, self-learning, modelling, simulation	Free
BUS++	Energy performance, ventilation, air flow, indoor air quality, noise level	
BV2	Annual energy use, durational diagram	
CELLAR	Cellar, heat loss, design rules	
CHP Capacity Optimizer	CHP, cogeneration, capacity optimization, distributed generation	Free
COMFIE	Energy performance, design, retrofit, residential buildings, commercial buildings, passive solar	
D-Gen PRO	Distributed power generation, on-site power generation, CHP, BCHP	
Demand Response Quick Assessment Tool	Demand response, load estimation, EnergyPlus	Free
DEROB-LTH	Energy performance, heating, cooling, thermal comfort, design	
DesiCalc	Desiccant system, air-conditioning, system design, energy analysis, dehumidification, desiccant-based air treatment	
Design Advisor	Whole-building, energy, comfort, natural ventilation, double-skin facade	Free
DesignBuilder	Building energy simulation, visualisation, CO_2 emissions, solar shading, natural ventilation, daylighting, comfort studies, CFD, HVAC simulation, pre-design, early-stage design, building energy code compliance checking, OpenGL EnergyPlus interface, building stock modelling, hourly weather data, heating and cooling equipment sizing	

Table 5.4.1.1. (Continued)

Tool	Applications	Cost
DOE-2	Energy performance, design, retrofit, research, residential and commercial buildings	
e-Bench	Energy benchmarking, environmental benchmarking, energy audit, invoice verification and reconciliation, performance contract verification	
EA-QUIP	Building modelling, energy savings analysis, retrofit optimization (work scope development), investment analysis	
ECOTECT	Environmental design, environmental analysis, conceptual design, validation; solar control, overshadowing, thermal design and analysis, heating and cooling loads, prevailing winds, natural and artificial lighting, life cycle assessment, life cycle costing, scheduling, geometric and statistical acoustic analysis	
EE4 CBIP	Whole building performance, building incentives	Free
EE4 CODE	Standards and code compliance, whole building energy performance	Free
EED	Earth energy, boreholes, ground heat storage, ground source heat pump system (GSHP)	
EN4M Energy in Commercial Buildings	Energy calculation, commercial buildings, bin method, economic analysis	
ENER-WIN	Energy performance, load calculation, energy simulation, commercial buildings, day lighting, life-cycle cost	
EnerCAD	Building energy efficiency, early design optimisation, architecture oriented, solar toolbox, SIA 380/1 compliance	
Energy Expert	Energy tracking, energy alerts, wireless monitoring	
Energy Profile Tool	Benchmarking, energy efficiency screening, end-use energy analysis, building performance analysis, utility programs	
Energy Scheming	Design, residential buildings, commercial buildings, energy efficiency, load calculations	
Energy Usage Forecasts	Degree days, historical weather, mean daily temperature, load calculation, energy simulation	
Energy-10	Conceptual design, residential buildings, small commercial buildings	
EnergyGauge USA	Residential, energy calculations, code compliance	
EnergyPlus	Energy simulation, load calculation, building performance, simulation, energy performance, heat balance, mass balance	Free
EnergyPro	California Title 24, compliance software, energy simulation, commercial, residential	
ENERPASS	Energy performance, design, residential and small commercial buildings	
ESP-r	Energy simulation, environmental performance, commercial buildings, residential buildings, visualisation, complex buildings and systems	Free
EZ Sim	Energy accounting, utility bills, calibration, retrofit, simulation	

Table 5.4.1.1. (Continued)

Tool	Applications	Cost
EZDOE	Energy performance, design, retrofit, research, residential and commercial buildings	
FEDS	multibuilding facilities, energy simulation, retrofit opportunities, life cycle costing, emissions impacts, alternative financing	
FLOVENT	Airflow, heat transfer, simulation, HVAC, ventilation	
FSEC 3.0	Energy performance, research, advanced cooling and dehumidification	
Gas Cooling Guide PRO	Gas cooling, hybrid HVAC systems	
Green Building Studio	Building information modelling, interoperability, energy performance, DOE-2, EnergyPlus, CAD	Free
HAMLab	Heat air and moisture, simulation laboratory, hygrothermal model, PDE model, ODE model, building and systems simulation, MatLab, SimuLink, FemLab, optimization	Free
HAP	Energy performance, load calculation, energy simulation, HVAC equipment sizing	
HEAT2	Heat transfer, 2D, dynamic, simulation	
HEED	Whole building simulation, energy efficient design, climate responsive design, energy costs, indoor air temperature	Free
Home Energy Saver	Internet-based energy simulation, residential buildings	Free
HOMER	Remote power, distributed generation, optimization, off-grid, grid-connected, stand-alone	Free
HOT2 XP	energy performance, design, residential buildings, energy simulation, passive solar	Free
HOT2000	energy performance, design, residential buildings, energy simulation, passive solar	Free
Hydronics Design Studio	hydronic heating, radiant heating, simulation, design, piping	
ION Enterprise	Energy management, power quality, power reliability, cost allocation	
ISE	Thermal model, building zone simulation, MatLab/SimuLink	Free
LESOCOOL	Airflow, passive cooling, energy simulation, mechanical ventilation	
LESOSAI	Heating energy, energy simulation, load calculation, standards	
MarketManager	Building energy modelling, design, retrofit	
Microflo	CFD, airflow, air quality, thermal performance	
Micropas6	Energy simulation, heating and cooling loads, residential buildings, code compliance, hourly	
ModEn	Object-oriented simulation, energy simulation, controls, energy audit, energy-saving, energy performance, dynamic simulation, research, education, heating, air conditioning	
NewQUICK	Passive simulation, load calculations, natural ventilation, evaporative cooling, energy analysis.	

Table 5.4.1.1. (Continued)

Tool	Applications	Cost
ParaSol	Solar protection, solar shading, windows, buildings, solar energy transmittance, solar heat gain coefficient, energy demand, heating, cooling, comfort, daylight	Free
Physibel	Heat transfer, mass transfer, radiation, convection, steady-state, transient, 2-D, 3-D	
PVcad	photovoltaic, facade, yield, electrical	Free
REM/Design	Energy simulation, residential buildings, code compliance, design, weatherization, equipment sizing, EPA Energy Star Home analysis	
REM/	Home energy rating systems, residential buildings, energy simulation, code compliance, design, weatherization, EPA Energy Star Home analysis, equipment sizing	
Right-Suite Residential for Windows	Residential loads calculations, duct sizing, energy analysis, HVAC equipment selection, system design	
RIUSKA	Energy calculation, heat loss calculation, system comparison, dimensioning, 3D-modelling	
RL5M	Residential, cooling, heating, energy, economic analysis.	
Room Air Conditioner Cost Estimator	Air conditioner, life-cycle cost, energy performance, residential buildings, energy savings	Free
SIMBAD Building and HVAC Toolbox	Transient simulation, control, integrated control, control performance, graphical simulation environment, modular, system analysis, HVAC	
SLAB	Slab on the ground, heat loss, design rules	
SMILE	Object-oriented simulation environment, building and plant simulation, complex energy systems, time continuous hybrid systems	Free
SMOG	Energy calculation, heat loss calculation, system comparison, dimensioning, 3D-modelling	
solacalc	Passive solar, house design, building design, building services, design tools	
SOLAR-5	Design, residential and small commercial buildings	Free
SolArch	Thermal performance calculation, solar architecture, residential buildings, design checklists	Free
SolarShoeBox	Direct gain, passive solar	Free
SPARK	Object-oriented, research, complex systems, energy performance, short time-step dynamics	Free
SUNDAY	Energy performance, residential and small commercial buildings	
SUNREL	Design, retrofit, research, residential buildings, small office buildings, energy simulation, passive solar	
System Analyzer	Energy analyses, load calculation, comparison of system and equipment alternatives	
TAS	Building dynamic thermal simulation, building simulation, comfort, CFD, thermal analysis, energy simulation	
TRACE 700	Energy performance, load calculation, HVAC equipment sizing, energy simulation, commercial buildings	

Table 5.4.1.1. (Continued)

Tool	Applications	Cost
TREAT	Weatherization auditing software, Home Performance with Energy Star auditing tool, retrofit, single family, multifamily residential, mobile homes, HERS ratings, load sizing.	
TRNSYS	Energy simulation, load calculation, building performance, simulation, research, energy performance, renewable energy, emerging technology	
tsbi3	Energy performance, design, retrofit, research, residential and commercial buildings, indoor climate	
VIP+	Energy performance, code compliance, design, research, residential and commercial buildings, costs, environmental sustainable	
VIPWEB	Energy performance, code compliance, design, research, residential and commercial buildings, costs, environmental sustainable	Free
VisualDOE	Energy, energy efficiency, energy performance, energy simulation, design, retrofit, research, residential and commercial buildings, simulation, HVAC, DOE-2	
WISE	Hygrothermal model, building simulation, MatLab/SimuLink Tool	Free

5.4.2. Indoor Air Quality Software

Table 5.4.2.1. Indoor Air Quality software

Tool	Applications	Cost
AIRPAK	Airflow modelling, contaminant transport, room air distribution, temperature and humidity distribution, thermal comfort, computational fluid dynamics (CFD)	
Apache	Thermal design, thermal analysis, energy simulation, dynamic simulation, system simulation	
BUS++	Energy performance, ventilation, air flow, indoor air quality, noise level	
COMIS	Multizone airflow, pollution transport	Free
CONTAM	Airflow analysis; building controls; contaminant dispersal; indoor air quality, multizone analysis, smoke control, smoke management, ventilation	Free
DesiCalc	Desiccant system, air-conditioning, system design, energy analysis, dehumidification, desiccant-based air treatment	
ESP-r	Energy simulation, environmental performance, commercial buildings, residential buildings, visualisation, complex buildings and systems	Free
FLOVENT	Airflow, heat transfer, simulation, HVAC, ventilation	
I-BEAM	Indoor air quality, IAQ education, IAQ management, energy and IAQ	Free

Table 5.4.2.1. (Continued)

Tool	Applications	Cost
IAQ-Tools	Indoor air quality, 'sick' buildings, contaminant source control, tracer gas,	
IDA Indoor Climate and Energy	Design, energy performance, thermal comfort, indoor air quality, commercial buildings	
Indoor Humidity Tools	Indoor air humidity, dryness, condensation	
LoopDA	Airflow analysis, indoor air quality, multizone analysis, natural ventilation	
Macromodel Assess. Resident. Concentrations Comb. Gen. Pollutants	indoor air quality, research	
Microflo	CFD, airflow, air quality, thermal performance	
ModEn	Object-oriented simulation, energy simulation, controls, energy-saving, energy performance, dynamic simulation, education, heating, air conditioning	
myupgrades.com	HVAC updates, HVAC equipment selection, energy savings, up-sell	
PHOENICS	Computational fluid dynamics, air pollution, smoke and fire, air flow	
Thermal Comfort	Thermal comfort calculation, comfort prediction, indoor environment	
VentAir 62	Ventilación design, ASHRAE Standard 62	

REFERENCES

[1] Schwab, M. and Simonson, C. Review of building energy simulation tools that include moisture storage in buildings materials and HVAC systems. 2004. Draft Report IEAQ Annex 41, Zurich.

[2] Wit, M. WAVO. A simulation model for the thermal and hygric performance of a building. 2007.

[3] Al-Rabghi, O. and D. Hittle. Energy simulation in buildings: overview and BLAST example. 2001. Energy Conversion and Management 42.13, 1623-1635.

[4] BSim. Danish Building and Urban Research. Visited May 6. http://www.dbur.dk/english/publishing/software/bsim/>.CHVAC Features. Updated April 2003. [Accessed December 2010].

[5] Elite Software. Visited May 6, 2003. http://www.elitesoft.com/web/hvacr/chvacx.html. [Accessed December 2010].

[6] DOE-2. Building Energy Analysis Simulation ResearchGroup. 2003. http://simulationresearch.lbl.gov/. [Accessed December 2010].

[7] EnergyPlus A New-Generation Building Energy Simulation Program. Updated June 4, 2003. U.S. Department of Energy. 2003. http://gundog.lbl.gov/EP/ep_main.html. [Accessed December 2010].

[8] E20-II HAP (Hourly Analysis Program) – 8760 Hour Load and Energy Analysis. 2003. Carrier. http://www.commercial.carrier.com/details/1,CLI1_DIV12_ETI496,00.html. [Accessed December 2010].

[9] A TRaNsient SYstems Simulation program. Updated July 24, 2003. Solar Energy Laboratory University of Wisconsin. http://sel.me.wisc.edu/trnsys/. [Accessed December 2010].

[10] Mendes, N., I. Ridley, R. Lamberts, P.C. Philippi and K. Budag. UMIDUS: A PC Program for the Prediction of Heat and Moisture Transfer in Porous Building Elements. 2003. http://www.hvac.okstate.edu /pdfs/bs99/papers/C-02.pdf. [Accessed December 2010].

[11] Presentation of the international building physics toolbox for SIMULINK Peter Weitzmann1, Angela Sasic Kalagasidis, Toke Rammer Nielsen1, Ruut Peuhkuri1 and Carl-Eric Hagentoft.

[12] Kalagasidis A.S. BFTools H Building physiscs toolbox block documentation. 2002.Department of Building Physics. Chalmer Institute of Technology. Sweeden.

[13] Kalagasidis A.S. HAM-Tools. International Building Physics Toolbox. Block documentation.

[14] Weitzmann P., Kalagasidis A.S., Nielsen T.R., Peuhkuri R., Hagentoft C. Presentation of the international building physics toolbox for simulink.

[15] Nielsen T.R., Peuhkuri R., Weitzmann P., Gudum C. 2002. Modelling Building Physics in Simulink. BYG DTU Sr-02-03. ISSN 1601-8605.

[16] Rode C., Gudum, Weitzmann P., Peuhkuri R., Nielsen T.R., Sasic Kalagasidis A., Hagentoft C-E. 2002. International Building Physics Toolbox-General Report. Department of Building Physics. Chalmer Institute of Technology. Sweden. Report R-02: 2002. 4.

[17] International Building Physics Toolbox. www.ibpt.org. [Accessed December 2010]

[18] Building Energy Software Tools Directory. Updated Feb. 7, 2003. U.S. Department of Energy. http://www.eere.energy.gov/ buildings/ tools_directory/. [Accessed December 2010]

SECTION 3: PRACTICAL CASE STUDIES

INTRODUCTION

In these last sections the reader will found four chapters that summarise the principal author research activities during this last years. The indoor ambiences will be analysed in these chapters were classified in accordance with the principals four groups of ambiences. In consequence we can find a first chapter about heath effect, others about energy saving and materials conservancy and a last chapter about work risk prevention in industrial environments.

Chapter 6

INDOOR AIR AND HEALTH EFFECTS: A PRACTICAL CASE STUDY OF FLATS

ABSTRACT

The present chapter reports results from an investigation involving a set of homesteads (flats) located in the area of Coruña, Spain, with the purpose of determining indoors comfort and health conditions.

In this particular case study, twenty five flats were chosen in such a way that at least one of the occupants have suffered or were suffering of a respiratory ailment at the time of the study. Experiments were carried out over 24 hour periods by keeping the household life as regular as possible. It was involved parameters such as temperature and humidity ratio, air renovations and microbiological load. In particular the air renovation was sampled from carbon dioxide concentrations that exist from the natural breathing process of its occupants.

Results have shown that, in general, comfort conditions were maintained in all the flats during the investigation and the ventilation rate was considered rather poor. As a result, the level of fungi and mesophiles presence found in some of the flats was relatively high.

6.1. INTRODUCTION

Located in the northwest coast of Spain, the climate of Coruña is mild. However, indoors humidity in the area is relatively high over most of the year due to the effect of the Atlantic Ocean winds. Health authorities claim that the level of incidence of respiratory ailments in the area is higher than in the rest of the country, a possible cause being the interior conditions of flats. In order to effectively confirm this claim, a systematic investigation of indoors conditions of homesteads in the area would be needed. This need prompted the research reported in the present chapter, in which data obtained in several flats in the Coruña area is presented and analyzed. The flats were randomly chosen and data was obtained under every day life of their occupants.

6.2. MATERIALS

During the investigation process it was employed some of the apparatus explained in the chapter 4 in accordance with the methodology of ISO Standard:

- A multi-gas monitor and a multi-sampler (Figure 4.3.1.1) were employed to sample the ventilation rate in different points at the same time. The apparatus was equipped with a temperature transducer to measure the state of the air at the point of sampling.
- Temperature and humidity were measured through tinytag dataloggers (Figure 4.4.1.1), which were adequately located so that a typical air condition of the room could be measured.
- A Casella AFC124 air suction pump was used in sampling air for microbiota analysis. The sample of air used to be filtered by flowing through a 47mm diameter, 0.45 µm pore ALBERT-NCS-045-47-BC cellulose nitrate membrane filter with an ALBERT PF-50-P-02 sterilized polycarbonate filter holder.

6.3. METHODS

6.3.1. Temperature and Humidity

As previously mentioned, the flats were randomly chosen so that typical every day life of occupants was not disturbed and typical indoor conditions could be obtained during the measurements. Indoor temperature and humidity have been measured in twenty five flats in the Coruña area. Measurements used to be taken over a period of 48 hours in each of the flats in accordance with chapter 4. Samples of outdoor air and air form the different rooms of the flat were also obtained during the measurement period. Indoor conditions have been referred to the ASHRAE Handbook of Fundamentals [1] showed in Table 1.1.1.

6.3.2. Thermal Comfort/Indoor Air Quality

Local thermal comfort has been evaluated in terms of the parameter PD, Percentage of Dissatisfied Persons, through the equation by Toftum et al (1988) and Simonson et al [2] and showed in Eq. 2.3.5.1. We must remember that an acceptable environment would be that in which less than 15% of the occupants are dissatisfied.

The Indoor Air quality has been evaluated through the so-called Acceptable Indoor Air Quality parameter, Acc with the Eq. 3.1.2.2 proposed by Fang et al. [3]. This equation showed that this parameter is strongly influenced by the temperature and the relative humidity, being linearly related to the air enthalpy.

Indoor Air Quality parameter is a measure of the level of acceptability of air with no known contaminants, as determined by a pertinent authority, and a level of dissatisfied occupants relatively small (lower than 20%).

6.3.3. Ventilation Rate

In this case, due to the existence of a natural CO_2 concentration in the flats, as a consequence of the occupant's breathings, it was selected as tracer gas for long periods of time. We must consider that another tracer gas like SF_6 must be employed for long periods and, in consequence will be more expensive besides results will be very similar to that obtained.

The ventilation rate has been determined in fifteen bedrooms in accordance with the chapter 3. In this case it was determined the ventilation rate through the concentration of carbon dioxide (CO_2) procedure. To do it, CO_2 was sampled with the gas monitor, despite the fact that it could be measured with the portable CO_2 datalogger (Figure 4.4.3.1) of chapter 4, in intervals varying between 11 and 14 minutes.

After the sample process, the air renovation and the minimum air renovation were obtained from Eq. 5.3.2.1.1 of chapter 3 but adapted to this case study. In consequence, we reach the Eq. 6.3.3.1.

$$\frac{dC_b}{dt} = \frac{N_0(C_{oa} - C_b) + F(t)}{V} \qquad (6.3.3.1)$$

where C_{oa} and C_b are the outdoor and bedroom CO_2 concentrations (ppm), V is the bedroom volume (dm^3), N_o is the natural ventilation rate, and F(t) is the CO_2 source (dm^3/s) in the room due to its occupancy.

In this case, the natural ventilation rate, N_o, can be determined from Eq. 6.3.3.1 given the outdoor and indoor CO_2 concentrations and the indoor CO_2 production. The minimum ventilation rate, N_{min}, is the one corresponding to a steady state CO_2 concentration with a 1000 ppm we can see in Eq.6.3.3.2 obtained from Eq. 3.2.1.2.

$$N_{min} = \frac{F(t)}{(C_{1000} - C_0)} \qquad (6.3.3.2)$$

6.3.4. Microbiological Load

A study of respiratory ailments was carried out during involving twenty-five family flats, and a total of 100 individuals. Flats were chosen in such a way that at least one of the occupants have suffered or were suffering of a respiratory ailment at the time of the study. Temperature and relative humidity was measured along with sampling of air to determine the microbiological load through culture and count.

Data gathering was complemented with questionnaires involving questions such as the state of the flat, living habits, indoor air quality perception, health and symptoms experienced by the occupants.

6.4. RESULTS AND DISCUSSION

6.4.1. Temperature and Humidity

The average indoor air temperature in most of the cases remained above 20ºC for an outdoor average temperature of 12ºC. Humidity has been evaluated in terms of the humidity ratio of the air. The relative behavior of the flats with respect to the humidity is shown in Figure 6.4.1.1. Two main conclusions can be drawn form this figure:

- The outdoor humidity is generally lower than indoors, a result that should be expected.
- The average humidity varied between 5.74 g/kg and 9.5 g/kg over the whole set of rooms and homesteads. These extremes have been obtained in the bedroom.

The average humidity ratio related to the whole set of flats have been determined for the kitchen, the living room, the bedroom and the kids bedroom for the following periods of time: breakfast, from 7:00 to 10:00, lunch, from 12:00 to 15:00, dinner, 20:00 to 21:00, and the sleeping time, 12:00 to 7:00. In addition, with exception of the kitchen at lunchtime, the humidity in the bedroom is the highest in the other periods of time.

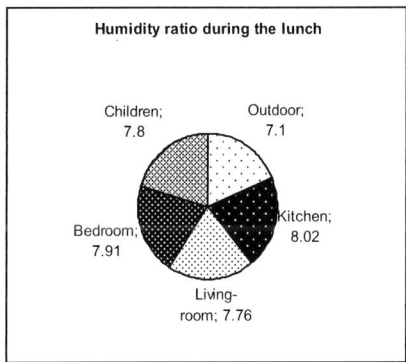

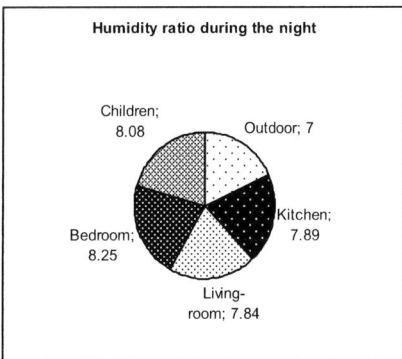

Figure 6.4.1.1. (Continued).

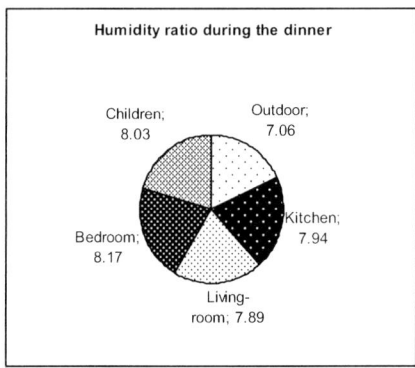

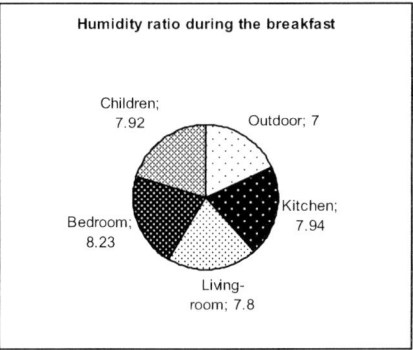

Figure 6.4.1.1. Humidity conditions in different zones of flats.

Finally, it was noticed that the range between 8 g/kg and 9 g/kg is by far the most frequent one. In fact, during the sleeping time, this range occurs almost 40% of the time in the bedroom. This range occurs 36% and 32% of the time at the breakfast and lunch times respectively in the kitchen. The range between 9 g/kg and 10 g/kg occurs 28% of the time in the kitchen during the dinnertime.

6.4.2. Comfort

Indoors comfort conditions have been evaluated by plotting indoor conditions in the perception of IAQ psychrometric chart, as shown in Figure 6.4.2.1, for the bedroom during the sleeping time and the kitchen during the lunch time, respectively. It can be noted that the relative humidity in the kitchen does not exceeds 60%, whereas for the bedroom this is slightly higher, reaching the order of 65%. Though these figures involving the relative humidity might be considered relatively high, it can be noticed that most of the conditions plotted in these figures are within acceptable limits with respect to the comfort and indoor air perception parameters, PD and Acc. In fact, the Acc parameter varies in the range from 0 to approximately 0.6 for most of the conditions. As for the parameter PD, most of the indoors conditions are within the range from 5% to 15%, with some of the conditions exceeding the 15% limit.

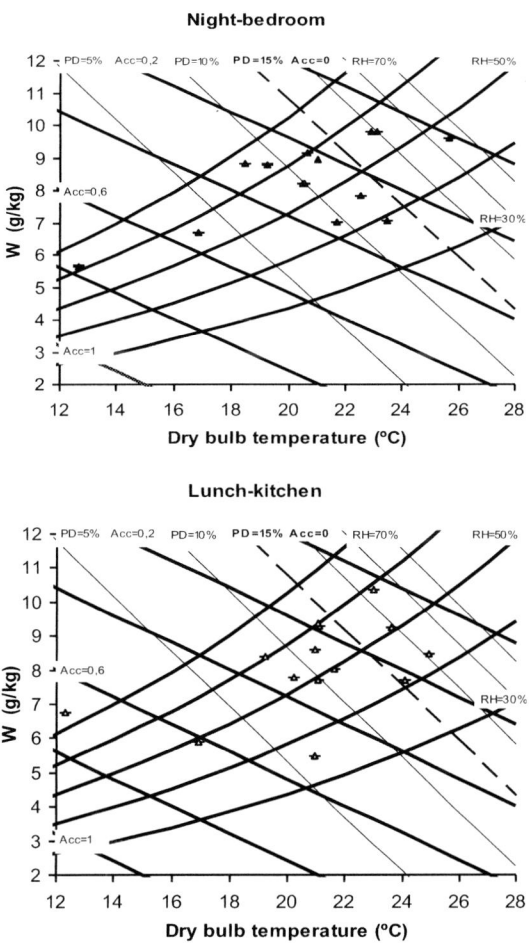

Figure 6.4.2.1. Acceptability and comfort conditions in kitchens and bedrooms.

6.4.3. Ventilation Rate

The ventilation rate is closely related to the CO_2 concentration in the room. It could be seen that the maximum concentration of CO_2 varied form a minimum of 2040 ppm to a maximum of 5480 ppm. A companion research involving public buildings (libraries, schools, etc) [4] revealed that the levels of CO_2 concentrations were lower than 800 ppm, indicating that the values for homesteads are abnormally elevated. In particular, CO_2 concentrations higher than 3000 ppm have been obtained in bedrooms with the door closed, even with one occupant. The average ventilation rate varied between 0.66 dm^3/s and 5.04 dm^3/s, being significantly lower than the minimum ventilation rate.

Finally, the effect of the relative ventilation between different rooms in the flats has been analyzed by comparing the humidity levels in different rooms through a variance study. The results showed that in most of flats exist a clear influence between conditions in the kitchen/living room couple during the dinnertime. Higher matching incidence might be associated to ventilation habits in the household. In fact, the high matching level observed at the dinnertime seems to point toward a significant air circulation between rooms in the flat. This circulation tends to promote a more uniform humidity between both environments, possibly being related to the habit of closing external doors and windows at that time. It is interesting to note that there has been no matching between the rooms of the flats during the sleeping time (night) and the morning.

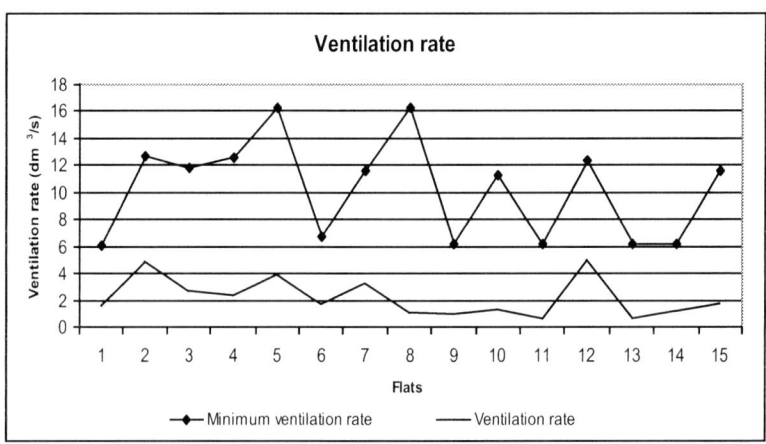

Figure 6.4.3.1. The calculated values and measurements during the sleeping interval for thirteen representative cases.

6.4.4. Microbiological Load

It has been found that the indoor humidity is closely related to the presence of microorganisms in the homesteads. Tables 6.4.4.1 and 6.4.4.2 present indoor conditions in terms of temperature and relative humidity and their ranges during the measuring period along with indices associated to the presence of microorganisms such as fungi and mesophiles. Table 6.4.4.1 only data for flats where the relative humidity during the measuring period remained below 60% has been considered. It can be noted that the fungi in the lower humidity flats are present in much lower levels. The overall fungi load does not seem to be related either to the number of occupants or to pets. There seems to be a direct relation between the damp on the walls and roofs and the fungi load. It has been found that 36% of the flats present dump problems in walls and roofs. On the other hand, the mesophiles, for the same temperature level, depend on the number of occupants, the presence of pets and the hygiene habits.

The microbiological load increases with the inadequate of thermal insulation. In flats where extensive areas are submitted to condensation due to poor thermal insulation, the development of fungi load has been observed. Finally, it has been determined that efficient cleaning procedures can contribute to improve indoor conditions and healthy environments due to the reduction in the microbiological production.

Table 6.4.4.1. Minimum, maximum and means values of measured parameters in buildings

N=25	Mean
Ind. Temp.	20.95
Ind. RH	60.40
Fungi	93.18
Mesophiles	548.96

Table 6.4.4.2. Minimum, maximum and means values of measured parameters in buildings with relative humidity below 60%.

N=11	Mean
Ind. Temp.	20.93
Ind. RH	55.72
Fungi	37.67
Mesophiles	564.90

CONCLUSIONS

In summarise we can say that the bedroom is the zone of the homestead where the humidity is higher, especially during the sleeping time. Values of the indoor humidity ratio found in the present investigation are generally higher than the outdoors and varied up to a maximum of 9 g/kg. Furthermore, It has been found that the natural ventilation rate in the considered flats was rather poor, reaching values significantly lower than those predicted by the minimum ventilation rate. As a result, the CO_2 concentrations in the flats were well above the levels found in public buildings.

The observed relative humidity in some of the rooms was found to attain values up to 65%, a level that could be considered relatively high. However the level of comfort given by the PD and Acc parameters for most of the observed indoors conditions remained acceptable during the measurement period.

- As expected should be expected, the ventilation rate tends to promote more uniform humidity and CO_2 content conditions in the flats.
- Control of the indoor humidity is important at temperatures around 21°C, typical for the present investigation, in preventing the development of fungi. It has been observed that fungi load diminishes in flats where the indoor relative humidity varied up to a maximum of 60% as compared with the results for flats in which the relative humidity could attain values even higher than the 60% level.
- The general conclusion to be drawn regarding air exchange is that given the current levels of occupation in the flats, ventilation procedures should be modified to keep the relative humidity lower than the maximum recommended of 60%. The natural ventilation can be effectively used in keeping both indoors temperature and humidity as well as the microbiological load within acceptable limits.

Finally, the author believes that information of the kind presented herein could be useful in the promotion of indoors healthy and comfort conditions in homesteads.

REFERENCES

[1] ASHRAE. Fundamentals. 1993. Atlanta. American Society of Heating, Refrigeration and Air Conditioning Engineers, Inc.

[2] Simonson C.J., Salonvaara M. and Ojalen T. Improving Indoor Climate and Comfort with Wooden Structures. 2001. Technical research centre of Finland. Espoo.

[3] Fang L., Wargocki P., Witterseh T., Clausen G. and Fanger P.O. Field study on the impact of temperature, humidity and ventilation on perceived air quality. 1999. The 8th International Conference on Indoor Air Quality and Climate. Edinburgh, Sotland 8–13, 2, 107.

[4] Rodriguez E., Baaliña A., Vazquez L., Castellanos J.A., Santaballa C.R. Infante, Indoor air quality evaluation using carbon dioxide levels in bedrooms in La Coruña (Spain). 1999. The 8th International Conference on Indoor Air Quality and Climate. Edinburgh, Sotland 8–13, 5, 335.

Chapter 7

INDOOR AIR AND ENERGY SAVING: A PRACTICAL CASE STUDY OF OFFICES AND SCHOOLS BUILDINGS

ABSTRACT

A Coruña, located in the north west of Spain, present a mild climate with a high relative humidity as a consequence of winds. Public buildings, like Spanish School, present the highest energy consumption in air conditioning during the winter season but during the spring the heating system is employed only if indoor conditions are under certain temperature and relative humidity values. Some energy saving methods could be employed to reduce or, in some cases substitute, the heating systems and, in consequence, reduce the energy consumption. Some of these methods are centred in an adequate design and application of HVAC system while others are focused in the study of heat and mass transfer through buildings envelops. A clear example of these methods was showed when permeable coverings [1, 2] were employed to reduce relative humidity peaks in offices buildings but exist other parameters, like thermal inertia, that could help us to compare indoor ambiences respect energy saving. In this sense, actual software like HAM tools could be employed to simulate indoor conditions and phenomena of material and energy transfer thought building envelopes and its effects on indoor conditions. Present chapter shows a study of methods to reduce the energy consumption in different kinds of schools buildings by means real sampled data and HAM tools simulations. Results showed that parameters like thermal inertia could be interfered principally by solar heat gains changing the buildings time constant. Other parameters like air changes per hour or permeable coverings present a clear enhancement of indoor ambiences.

7.1. INTRODUCTION

A Coruña, located in the north west of Spain, present a mild climate with a high relative humidity as a consequence of winds. Public buildings, like Spanish School, present the highest energy consumption in air conditioning during the winter season but during the spring the heating system is employed only if indoor conditions are under certain temperature and relative humidity values. Some energy saving methods could be employed to reduce or, in some cases substitute, the heating systems and, in consequence, reduce the energy consumption. Some of these methods are centred in an adequate design and application of HVAC system while others are focused in the study of heat and mass transfer through buildings envelops. A clear example of these methods was showed when permeable coverings [1, 2] were employed to reduce relative humidity peaks in offices buildings but exist other parameters, like thermal inertia, that could help us to compare indoor ambiences respect energy saving.

The system with the highest thermal inertia has the lowest energy requirement because in times of abundance (due to solar irradiation…) energy is stored in internal and external building constructions. This energy is transferred back into the zone when the indoor temperature decreases, thus limiting heat needed from the heating system without significantly affecting the thermal comfort [3]. The fact that a high time constant for the thermal processes in a building decreases the energy requirement for heating makes it possible to choose a higher design winter temperature or change the working conditions of the HVAC system like a change from intermittently controlled heat pumps to continuously capacity controlled heat pumps [4]. Other investigations are centred in characterization of thermal inertia as a part of the installation of a system of summer refreshment by means of nighttimes cooling ventilation [5] and others in the study passive solar buildings [6, 7, 8].

To understand and predict these effects software tools were employed but underestimated the energy consumption of buildings by up to 60% because the energy models ignore moisture. In this sense, whole building performance can only be realistically evaluated by accounting for the HAM interactions. Through its Energy Conservation in Buildings and Community Systems

Program, the International Energy Agency launched Annexes 17 and 41 [9], a working group to address issues surrounding whole building HAM response. The group involved over 50 researchers from 28 institutes and over 20 countries as this Department of Energy [10]. This Annex 41 tested Building simulation software like H-tools and HAM Tools from Chalmer Institute of Technology to simulate heat and mass transfer through buildings envelopes. These tools are employed to simulate indoor conditions as phenomena of material and energy transfer thought building envelopes and its effects on indoor conditions considering heat gains like occupation, solar heat, illumination and air renovation and infiltrations between others. Once this software is being tested new conclusions could be obtained. In present chapter a study about the methods to reduce the energy consumption in different kinds of schools buildings by means real sampled data and HAM tools simulations is done.

7.2. MATERIALS

In this case, due to the fact that we want sample indoor conditions for long periods of time, temperature and relative humidity were measured using geminy data logger (Figure 6.4.1.1), air renovation was measured with a multi-gas monitor (Figure 6.3.1.1) using the concentration decay method of SF_6.

7.3. METHODS

As the objective of this study is to find indoor ambience improvements that let us energy saving in school buildings, two school buildings were sampled during different seasons to relate indoor conditions with other parameters like weather, and heat and moisture balances. In this sense, ASHRAE Standard 1992 [11] indications and Burch [12, 13] simulations showed that the massiveness of an exterior wall reduces the heating and cooling requirements of buildings, provided the room air temperature floats above the thermostat set point in heating and below the thermostat in the cooling season. The floating temperature occurs more frequently in mild climates and during the spring. Special interest presents the spring season because then the HVAC system could be fully substitute by some passive methods. In consequence, this paper will study thermal inertia effect of two buildings with high and low wall density during the spring season of the mild weather of A Coruña (Spain).

The week period selected for this study was a weekend and a holiday to identify the appropriate indexes of thermal stability for building envelopes. The focus must consider the solar heat gain and heat storage of building walls under conditions of natural ventilation [14]. During this unoccupied period, occupation and excessive air renovations will not interferes the sampled data and an easy environment simulation could be done.

After tests the simulations indoor conditions modifications will be proposed for energy saving. In this sense, location of the mass in relation to the insulation has a large effect on the deviation between measured energy use and steady state analysis. In high-density buildings where mass was outside the insulation, the measured energy use closely matched that predicted by steady-state analysis but not when the insulation was outside the mass [11]. In consequence, parameters like thermal inertia, air renovation and internal coverings were simulated showing a clear different behaviour in each condition.

Conclusions let us understand if the passive methods that could let us reach better indoor condition during the first hour of occupation of the morning and evening class periods.

7.3.1. Schools

Two schools are sampled and simulated. One of the areas of the older school was built in 1890, and the other part was built in 1960 and the new school was built in 1999 as we can see in Figures 7.3.1.1 and 7.3.1.2. In consequence, the old school presents 0.43 m of stone and 0.5 cm of concrete in the indoor side of the wall. The wall of the new building consist in layers of insulation, brick, concrete and plaster arranged symmetric respect the middle of the wall and reaching 0.30 m of total thickness as we can see in Figure 7.3.1.3. The classroom sampled, in the old building, is located on the second floor and has a volume of 210 m^3, while the new is located on the first floor with a volume of 150 m^3.

All these buildings present a working period from February to June and an unoccupied period during the weekends and holidays. In those periods classrooms are under natural ventilation and central heating system was not employed. They active period ends in June and, in consequence, it is not interesting for energy saving during summer period. Furthermore, during the winter extreme conditions these schools are not working and, inconsequence heating system will works only when the indoor conditions exceed the thermal comfort during winter and spring.

Figures 7.3.1.1 and 7.3.1.2. Old and new school building of A Coruña respectively.

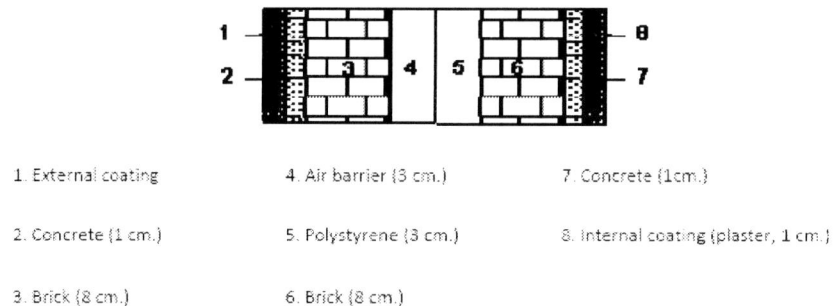

1. External coating
2. Concrete (1 cm.)
3. Brick (8 cm.)
4. Air barrier (3 cm.)
5. Polystyrene (3 cm.)
6. Brick (8 cm.)
7. Concrete (1cm.)
8. Internal coating (plaster, 1 cm.)

Figure 7.3.1.3. New school wall structure.

7.3.2. Indoor and Outdoor Sampling Conditions

The indoor and outdoor humidity and temperature have been monitored simultaneously in the most representative classroom of each school during part of winter and spring seasons. All schools have purely adventitious ventilation. Transducers were hung in the middle of the classrooms. Data has been gathered in Tiny tags data loggers which can store 7,600 readings.

7.3.3. Thermal Comfort and Indoor Air Quality Indexes

Local thermal comfort has been evaluated in terms of the parameter PD, Percentage of Dissatisfied Persons, through the Eq. 2.3.5.1 obtained by Toftum et al (1988) and Simonson et al [15-18], as was explained in chapter 2.

7.3.4. Time Constant

The time constant is normally found from a slow cooling down period with a constant low outdoor temperature as (heat capacity)/(heat loss factor) [3]. This method is based on a seasonal steady state energy balance on the building as a whole or on a particular building zone. The thermal inertia is introduced in terms of the utilisation factor that shows the part of energy gains (solar irradiation and others) that can be stored in building construction to be transmitted into the zone when needed, as we can see in Eq. 7.3.4.1.

$$Q_{heat} = Q_{loss} - \eta Q_{gain} \tag{7.3.4.1}$$

where

Q_{heat} is the heat requirement (W)
Q_{loss} is the heat loss (W)
Q_{gain} is the heat gain (W)
η is the utilisation factor, having a value between 0 and 1.

The utilisation factor η is a function of the building periodic time constant and the ratio Q_{gain}/Q_{loss}. The time constant is defined in the standard by the Eq. 7.3.4.2.

$$\tau = \frac{\sum C}{\sum H} \tag{7.3.4.2}$$

where;
C is the sum of thermal capacity C of each construction based on a 24 hour periodic response.
H is the sum of heat loss factor of each construction, ventilation and air leakage.

As [3] recommends, when we want to work in a more precise way, the logarithm of the temperature difference in-outdoors is taken and matched to a straight line by the method of least squares. The time constant is the inverse of the coefficient for the independent variable (time) given by this curve fit. In consequence, after test our simulations with real sampled data; both buildings were simulated under constant weather conditions with the aim to determine building time constants.

7.4. RESULTS

7.4.1. Thermal Inertia and Solar Gain: Time Constant Determination

Figures 7.4.1.1 and 7.4.1.2 represent the logarithm of the temperature difference between indoor and outdoor temperatures when building is under constant weather conditions. Its linear regression constants will give us the time constant of each building.

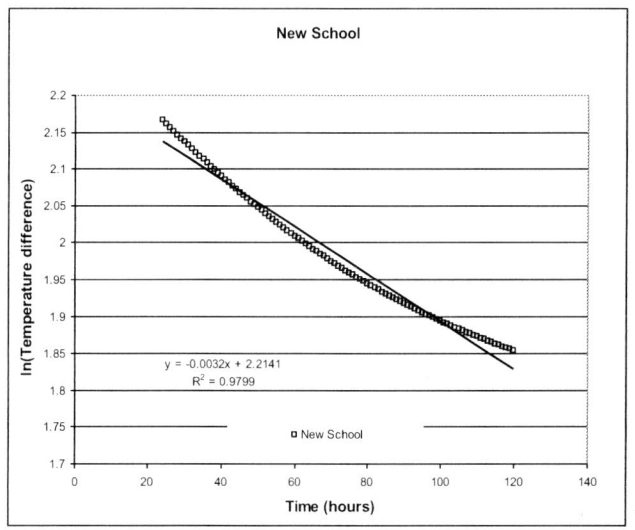

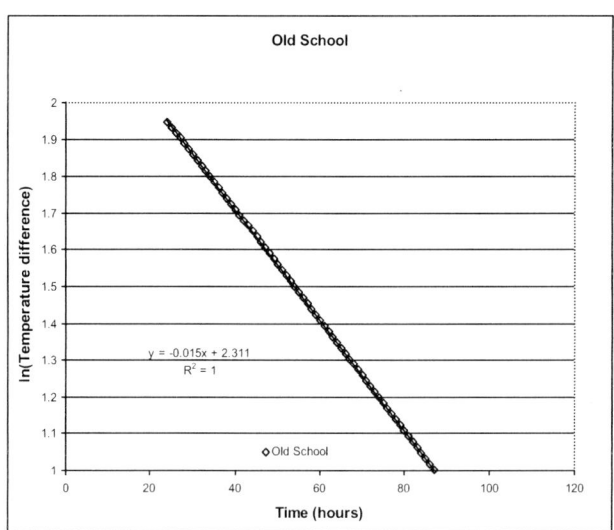

Figure 7.4.1.1. Time constant determination for new and old school buildings.

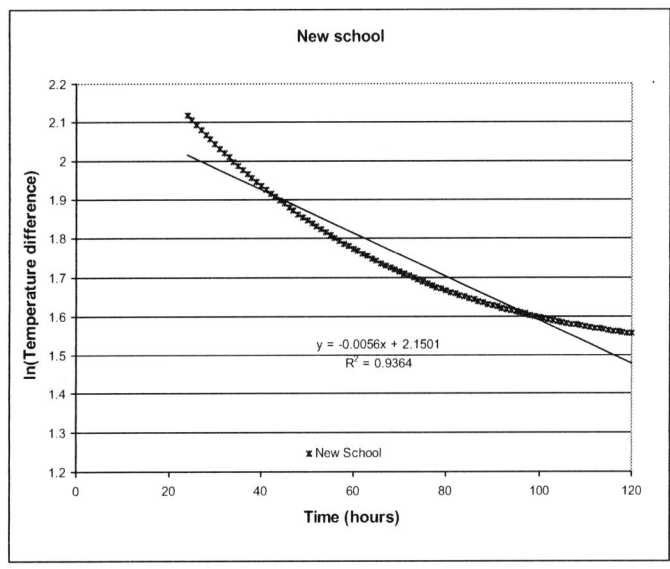

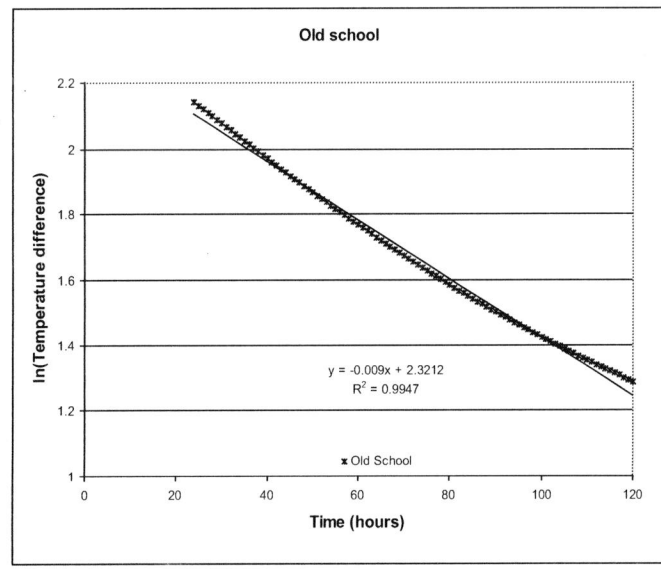

Figure 7.4.1.2. Time constant determination for new and old school buildings without heat gains.

7.4.2. Air Renovation

Sampled air renovations were changed from 0.7 during the unoccupied period for the old school and 0.6 in the new school to a lower value of 0.4 air renovations with the aim of observant the effect of weather on indoor conditions. Temperature and Percentage of dissatisfied are showed in Figures 7.4.2.1 and 7.4.2.2.

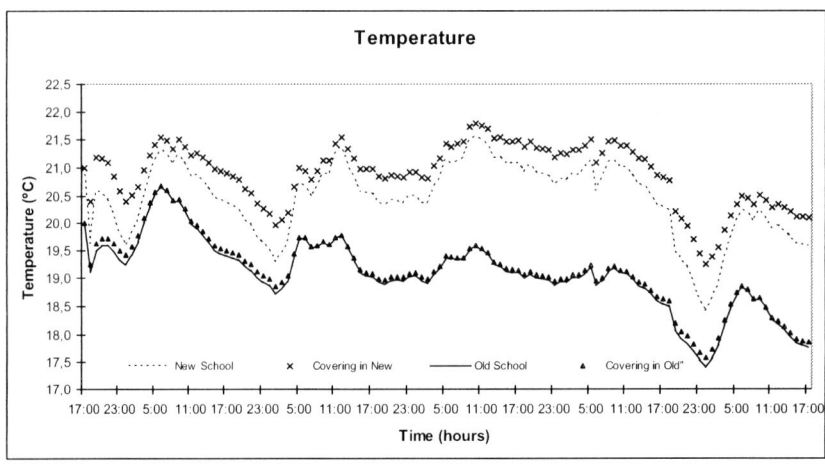

Figure 7.4.2.1. Temperature when air renovations were reduced.

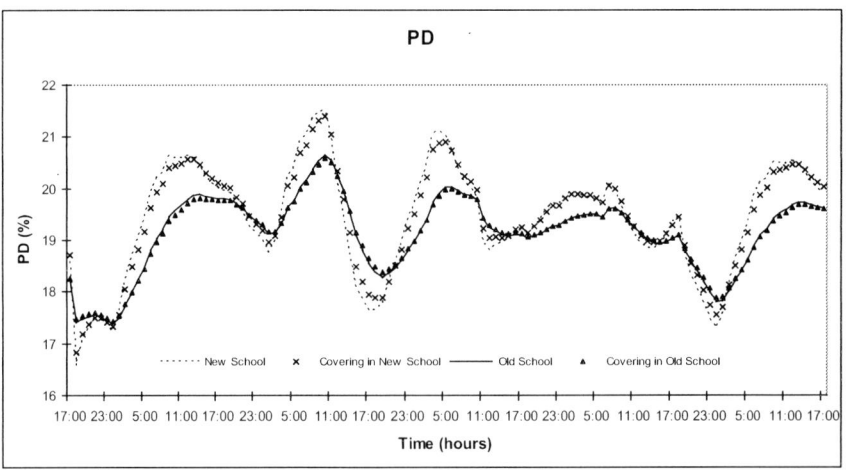

Figure 7.4.2.2. PD when air renovations were reduced.

7.4.3. Heat and Mass Transfer through Building Envelope

Now, the internal covering was changed in both buildings with the aim to improve indoor conditions. The permeable covering selected to substitute the concrete internal covering was 1.5 cm of wooden panel that could control the high indoor relative humidity, especially in the old school building. Results are showed in Figures 7.4.3.1 and 7.4.3.2.

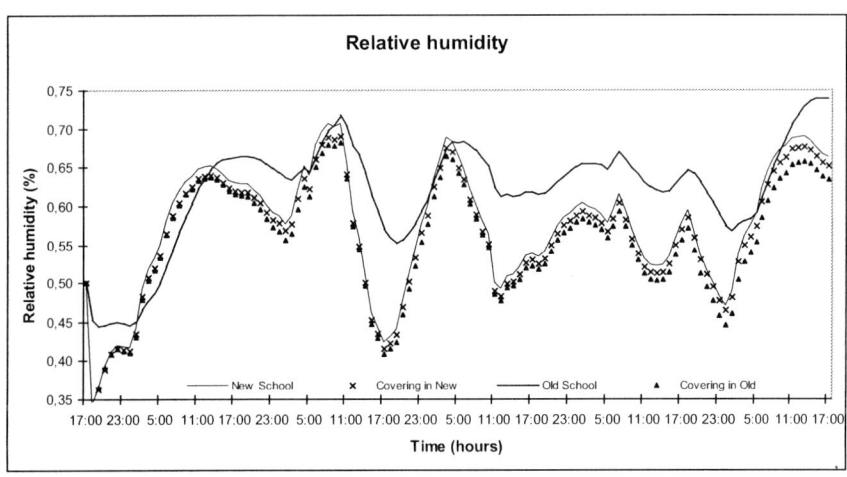

Figure 7.4.3.1. Relative humidity with and without internal coverings.

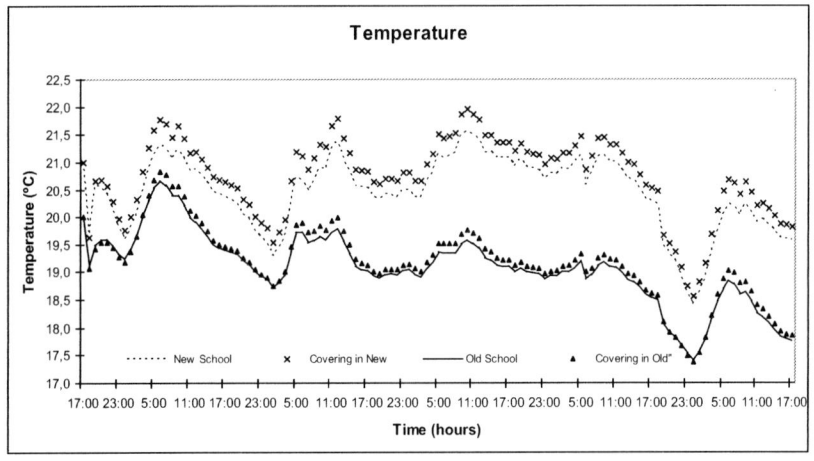

Figure 7.4.3.2. Temperature with and without internal coverings.

7.5. Discussion

Results of Figure 7.4.1.1 showed a time constant of 111 for the old and 178 in the new school with an adequate correlation factor in each case. This value shows a higher thermal inertia of the new school than the old. Other simulations were done under different indoor ambience temperatures obtaining the same value. The explanation of this effect is related with solar heat gains that the new building experiments respect the old as a consequence of the classroom way and the presence of another nearer buildings that interfere in that heat gain. To understand this solar effect, this same process was simulated but without this heat gain and another time constants were obtained in Figure 7.4.1.2. Once again adequate linear regressions were obtained with correlation values of 0.98 and 1 and time constants of 36.9 and 66.6 for new and old schools respectively. Now the old school present a higher thermal inertia respect the new as a consequence of the purely effect of wall thickness and heat transmission properties.

Once obtained these results other changes were proposed. For example, old school presents a high air renovation during the unoccupied period as a consequence of the air leakages. These air renovations reach values upper the 0.5 usually obtained in closed ambiences. When the air renovations of the two buildings are changed to 0.4 another curves are showed in the Figures 7.4.2.1 and 7.4.2.2. Indoor air humidity, in the new and old school, experiments a decrease to more adequate values of 60%. This effect is related with the heat and moisture transfer through stonewalls. The indoor air temperature reaches the same maximum values but experiment a slowly decrease in the new building reaching a higher minimum values than under normal conditions of air renovation. This thermal effect will be present in the indoor air enthalpy and, in consequence, on the percentage of dissatisfied that reaches slight lower PD maximum values during the night.

Finally, other internal coverings were proposed to change the indoor ambience conditions. In consequence, wooden panel was simulated as internal coverings in the two cases. Results are showed in Figures 7.4.3.1 and 7.4.3.2. In this sense, permeable coverings have result an adequate covering to control the extreme relative humidity evolution, especially in the old building. Now, we can see that indoor relative humidity was extremely changed in the old school reaching a value nearer the new building, which not experimented an apparent change in their indoor conditions. Indoor temperature experiments a clear enhancement in the new building reaching values of 1 °C higher than in the previous case. Less intense is this effect on the old buildings. As a consequence of these changes in indoor air temperature and relative humidity the percentage of dissatisfied reach nearly the same values than that obtained in normal conditions in old and in new schools. This is a consequence of the slightly changes in indoor air enthalpy, which is nearly constant when the proportion 1 °C-5% of relative humidity was maintained. For example, if indoor air temperature experiment an enhancement of 1 °C, relative humidity will experiment a decrement of 5%, in consequence, will exist nearly the same indoor air enthalpy and percentage of dissatisfied.

In this last simulation, despite the fact that exist a permeable covering, there was not an intense effect on indoor ambience. This is related with the reduce surface of permeable covering, which represent only the surface of the class to interact with the outdoor ambience, and the high volume of the old classroom (210 m^3).

The enthalpy conditions indicate that HVAC system is not needed because the indoor air enthalpy under natural ventilation reaches the value of 39 kJ/kg, estimated for adequate indoor conditions.

In resume, we can conclude that a reduction in indoor air renovations lead to the new building, where the solar heat gain is more important, to increment its indoor temperature and, in consequence, doing an indoor ambience more insensible to outdoor weather change, as we can see in Figure 9 in the indoor temperature slope after a peak of temperature.

On the other hand, when a permeable covering is proposed the temperature decay is parallel and slight higher than initial conditions enhancing a reduction on indoor air relative humidity. This increment of indoor air temperature is related with the insulation properties of the wooden panel.

As we have note, the importance of this time constants is based on the fact that the building with the lower time constant reacts faster on weather changes and variation in internal heat gains than the more heavy buildings. Even during a short period of cold weather heat must be supplied in the lightweight building, whereas such periods can be passed without heating in the constructions with higher thermal inertia due to heat stored in the structure from previous warmer periods in accordance with [3]. Furthermore, the amplitude of temperature fluctuation of the inner surfaces of walls made of low time constant buildings under intermittent air-conditioning conditions is 1 °C higher than that of walls of buildings with a higher thermal inertia under continuous air-conditioning conditions in accordance with [14].

These effects can be summarised in their study of time constant of each variant. In our case, when indoor air renovations are reduced, the time constant for new and old schools is 185 and 112 respectively, as we can see in the Table 7.5.1.

Table 7.5.1. Time constant for each modification

	Without heat gain	Initial conditions	Air renovation reduction	Permeable coverings
New	37	178	185	188
Old	67	111	112	115

These values are similar to that of 325, 164 and 31 for heavyweight, massive wood and lightweight walls showed by [3]. Despite this, exist an internal gain from convective heat source on new building that lead to a higher effect than the wall construction thermal inertia. In consequence, the energy saving and HVAC system design must be done in accordance with its individual characteristics of each building and not only taking into account meteorological data and general procedures.

CONCLUSIONS

This chapter sample and simulate different indoor conditions in buildings with the aim to determinate the possibility of energy saving and design considerations on building and HVAC systems.

The heat-transfer through the night thermal inertia elements is analysed by using 1D time dependent conduction heat transfer equation that is solved numerically by using HAM tools. The model takes into account in a detailed fashion the inertial heat sources and the air renovation. These simulations showed that old building presents the lowest thermal inertia than the new as a consequence of a solar heat gain. In consequence, the building with the lower time constant reacts faster on weather changes and variation in internal heat gains than the more heavy buildings. Parameters like air renovation and permeable coverings interact in over the time constant. If the air renovation is reduced the old building experiment a slowly increment, while the new will experiment a clear internal temperature increment as a consequence of the heat gain. Finally, the presence of permeable internal coverings like wood panel let us increment the time constant as a consequence of the increment of wall insulation, especially in the old school.

The effect of permeable coverings over the indoor ambience was reduced under the presence of a little surface of permeable coverings. In consequence, more research is needed to define new design and corrections of HVAC systems taking into account these individual parameters of each building location.

REFERENCES

[1] Orosa J.A., Baaliña A. Passive climate control in Spanish office buildings for long periods of time. 2008. Building and Environment 2008; doi:10.1016/j.buildenv.2007.12.001

[2] Orosa J.A., Baaliña A. Improving PAQ and comfort conditions in Spanish office buildings with passive climate control. 2008. Building and Environment 2008; doi:10.1016/j.buildenv.2008.04.013.

[3] Norén A., Akander J., Isfält E., Söderström O., The effect of Thermal Inertia on Energy Requirement in a Swedish Building-Results Obtained with Three Calculation Models. 1999. International Journal of Low Energy and Sustainable Buildings. 1.

[4] Karlsson F., Fahlén P. Impact of design and thermal inertia on the energy saving potential of capacity controlled heat pump heating systems. 2008. International Journal of refrigeration. 31, 1094-1103.

[5] Roucoult J.M., Douzane O., Langlet T. Incorporation of thermal inertia in the aim of installing a natural night time ventilation system in buildings. 1999. Energy and Buildings. 29, 129-133.

[6] Badescu V., Sicre B. Renewable energy for passive house heating II Model. 2003. Energy and Buildings. 35, 1085-1096.

[7] Badescu V., Sicre B. Renewable energy for passive house heating Part I. Building description. 2003. Energy and Buildings. 35, 1077–1084.

[8] Krüger E., Givoni B. Thermal monitoring and indoor temperature predictions in a passive solar building in an arid environment. 2008. Building and Environment. 43, 1792-1804.

[9] Hauer A., Mehling H., Schossig P., Yamaha M., Cabeza L., Martin V., Setterwall F. International Energy Agency Implementing Agreement on Energy Conservation through Energy Storage. Annex 17. "Advanced Thermal Energy Storage through Phase Change Materials and Chemical Reactions – Feasibility Studies and Demonstration projects". Final Report.

[10] International Energy Agency. http://www.iea.org. [Accessed December 2010]
[11] ASHRAE Handbook Fundamentals. Load and Energy Calculations, Energy Estimating Methods. 1993. Chap. 28.
[12] Burch M.D., Chi J. MOIST A PC Program for Predicting Heat and Moisture Transfer in building Envelopes. 1997. NIST Special Publication 917. NIST United States Department of Commerce Technology Administration. National Institute of Standards and Technology.
[13] Burch D.M., Remmert W.E, Krintz D.F. and Barnes C.S. A Field Study of the Effect of Wall Mass on the Heating and Cooling Loads of Residential Buildings. National Bureau of Standards Washington, D.C. 20234. Proceedings of the Building Thermal Mass Seminar. Knoxville, TN; 6/2-3/82. Oak Ridge National Laboratory.
[14] Feng Y., Thermal design standard for energy efficiency of residential buildings in hot summer/cold winter zones. 2004. Energy and Buildings. 36, 1309-1312.
[15] Simonson C.J., Salonvaara M., Ojalen T. The effect of structures on indoor humidity-possibility to improve comfort and perceived air quality. 2002. Indoor Air. 12, 243-251.
[16] Simonson C.J., Salonvaara M., Ojalen T. Improving indoor climate and comfort with wooden structures. Espoo 2001. Technical Research Centre of Finland. VTT Publications. 431. 200+ app 91.
[17] Toftum J., Jorgensen A.S., Fanger P.O. Upper limits for indoor air humidity to avoid uncomfortably humid skin. 1998. Energy and Buildings. 28. 1-13.
[18] Toftum J., Jorgensen A.S., Fanger P.O. Upper limits of air humidity for preventing warm respiratory discomfort. 1998. Energy and Buildings. 28, 15-23.
[19] Kalagasidis A.S. BFTools H Building Physiscs Toolbox block documentation. Department of Building Physics. 2002. Chalmer Institute of Technology. Sweeden.
[20] Kalagasidis A.S. HAM-Tools. International Building Physics Toolbox. Block documentation.
[21] Weitzmann P., Kalagasidis A. S., Nielsen T. R., Peuhkuri R., Hagentoft C. Presentation of the international building physics toolbox for Simulink.

[22] Nielsen T.R., Peuhkuri R., Weitzmann P., Gudum C. Modelling Building Physics in Simulink. 2002. BYG DTU Sr-02-03. ISSN 1601-8605.
[23] Rode C., Gudum, Weitzmann P., Peuhkuri R., Nielsen T.R., Sasic Kalagasidis A., Hagentoft C-E. International Building Physics Toolbox-General Report. 2002. Department of Building Physics. Chalmer Institute of Technology. Sweden. Report R-02: 2002. 4.
[24] International Building Physics Toolbox www.ibpt.org. [Accessed December 2010]
[25] Wit, M. WAVO. A simulation model for the thermal and hygric performance of a building. 2000. Faculteit bouwkunde, Technische Universiteit Eindhoven.

Chapter 8

INDOOR AIR AND MATERIALS CONSERVANCY: A PRACTICAL CASE STUDY OF LIBRARIES

ABSTRACT

ASHRAE Standard defines naturally conditioned spaces as those where thermal conditions are regulated primarily by the occupants through opening and closing of the windows. It is known that non-domestic naturally ventilated buildings are likely to consume less energy than those that are air-conditioned which is related with the fact that this ambiences tend to make more effective use of natural light and partly because the electrical energy consume by fans, chillers and pumps is avoided. In this sense, naturally ventilated buildings can have low energy demands but the strategy is difficult to implement in urban locations. Present chapter shows a practical case study about indoor condition in typical Spanish library respect ASHRAE indications. Results showed temperature and relative humidity values higher than expected. In consequence a higher percentage of dissatisfied and materials deterioration was found. Passive climate control methods were proposed to get a sustainable building.

8.1. INTRODUCTION

Nowadays is known that non-domestic naturally ventilated buildings are likely to consume less energy than those that are air-conditioned. It is a consequence of the fact that this ambiences tend to make more effective use of natural light and partly because the electrical energy consume by fans, chillers and pumps is avoided. In this sense, naturally ventilated buildings can have low energy demands but the strategy is difficult to implement in deep plan, urban locations. For example, some natural ventilated libraries tend to employ passive cooling strategies [1-2], include in maritime ambiences [3]. In this sense, a mechanical ventilation system is employed to control both daily latent loads as well as air quality during the day whilst also facilitating cooling of the building during the night to get an indoor air temperature of 19 ºC at the beginning of occupancy the following day. As a consequence of this building sensible cooling, a peak reduction was observed in the afternoon period, when internal occupancy gains were highest.

Control of thermal comfort in buildings with centralized HVAC system tend to simulate real world systems with ESP-r to help improve and control its air-conditioning systems [3-5]. On the other hand, in naturally conditioned spaces thermal conditions are regulated primarily by the occupants through opening and closing of the windows as is defined by the ASHRAE Standard 55-2004. This same standard reflects the fact that occupant's thermal response in such spaces depends in part on the outdoor climate, and may be different in buildings with centralized HVAC system. This difference is due to different thermal experiences, changes in clothing, availability of control, and shifts in occupant expectations. Mechanical ventilation with unconditioned air may be utilized, but opening and closing of windows must be the primary means of regulating the thermal conditions in the space. For this spaces this standard shows an optional method to apply. The method of adaptive model is based on climatic data and presents an adaptive comfort ranging from 19 ºC to 24 ºC of internal temperature on any day, which is considered during the building design in [3]. It is not the only parameter in natural ventilated buildings. In the past, environmental control was oriented towards the convenience of the occupants but contemporary researches showed the danger of deterioration of materials [6]. These researches concluded that the three categories of materials deterioration in this kind of ambiences are; change in the size and the shape of the exhibits, changes in the rate of the deterioration chemical reactions, changes in biological deterioration sources. These groups of materials deterioration can be associated respect three factors; relative humidity, temperature, and indoor pollution. The first is strongly related with preservation of the materials in archives, libraries and museums. In this sense, ASHRAE Handbook 2003 HVAC Applications [7] recommend values for relative humidity and temperature but these values depend on the specific material. In particular, in libraries is especially interesting the change in the rate of deterioration by chemical reactions. Recent researches concluded that exist decrease of paper resistance and fading and the alteration of dye materials under a relative humidity of 60%. On the other hand, when the majority of museums exhibit textiles material, relative humidity measures shall be about 45% and music and films stored in closed stacks must be at 15.5 ºC and a relative humidity of 50%.

If the relative humidity at the surface is above a critical value of 70%, especially in hot environment with static air, fungal grow will be present on most surfaces. The growth rate depends on the magnitude and the duration of surface relative humidity. This surface relative humidity is a complex function of material moisture content, local surface temperature and humidity conditions in the space. In recognition of the issue's complexity, the International Energy Agency established a surface relative humidity criterion for design purposes: monthly average values should remain below 80%. Other proposals include the Canada Mortgage and Housing Corporation's stringent requirement of always keeping the surface relative humidity below 65%. Although there still is no agreement on which criterion is more appropriate, mould growth can usually be avoided by allowing surface relative humidity over 80% only for short time periods. This relative humidity criterion may be relaxed for nonporous surfaces that are regularly cleaned.

In the case of books, fungal colonisation causes significant disfigurement of library materials. Apart from the undesirable aesthetic aspect, the presence of some microorganisms can lead to the irreversible degradation of the paper. The environmental factors, which determine germination and growth on paper-based material, are the relative humidity, temperature, gas composition and the level of initial contamination of the material. Recent studies [8] showed that electronic nose technology could be used to effective monitoring in a storage situation, especially if linked to a real time neural network that has information on the volatile patterns from non-spoiled paper material. These results showed so that dust accumulation may be an important factor in providing foci for fungal activity and that relative humidity affects the volatile emission and the growth rate of the different fungal species.

Finally, moisture absorption and wetting cause visible and invisible degradation. This visible degradation includes mould surfaces, decay of wood-based materials, spalling of masonry and concrete caused by freeze-thaw cycles, hydration of plastic materials, corrosion of metals, damage from expansion materials and decline in appearance.

Temperature is the second parameter to consider. As was explained, temperature is related with the acceleration of chemical process of deterioration of materials. For example, it was shown that there exists an acceleration of the corrosion rate of cellulose when temperature reaches values of 20 °C. Furthermore, the partial drying up of books may result in fragility levels, especially if humidity is not kept stable [6]. A possible solution to solve this problem is the mass deacidification. It is a process involving the neutralisation of the acids present in the paper and the deposition of an alkaline buffer to prevent, or at least retard further acidification [9]. Results showed that after accelerate ageing methods, like humidity atmosphere, not only the treated paper exhibited higher folding endurance compared with its untreated aged counterpart, generally it led to the largest improvement in the mechanical properties upon ageing when it is compared with the unchanged stage. Furthermore, it can be assume that exist an effect at the room temperature, upon long-term storage.

Temperature recommendations for areas used exclusively for storage are much lower than those for combination user and storage areas. In general, the higher the temperature the more rapidly materials will deteriorate. However, because people must have access to library and archival materials, there is a practical limit to how low the temperature can be maintained. Ideally, the temperature should be closely controlled at 20 °C to 21 °C and consistently maintained in all areas of a storage facility. This temperature range is regarded as tolerable to staff and users and appropriate for most materials.

Indoor pollution is the third factor to consider. In general, indoor pollution in cultural institutions such as museums, libraries and archives is of particular importance because it is related with a healthy indoor climate and cultural assets protection against deterioration [10]. In this sense, results of recent investigations reaching to the conclusion that most of the detected VOCs are associated with packaging and building products used for furnishing magazines and building exhibition drawers.

In summarize, authorities disagree on the ideal temperature and relative humidity for library and archival materials. A frequent recommendation is a stable temperature no higher than 21 °C and a stable relative humidity between a minimum of 30% and a maximum of 50%. Finally, all library staff has a responsibility to be aware of acceptable environmental conditions for library materials. Library Department Heads and those assigned monitoring responsibility must be the first line of defence against problems involving temperature and humidity [11].

8.2. MATERIALS

8.2.1. Building

Library object of this study is principally occupied by University students but, besides that no relics are stored, the goal is to maintain within the specified ranges of temperature and humidity and minimize daily and weekly variations. This building does not have any other near build that interfere in the thermal ambience and presents two zones clearly differenced. The first zone corresponds with the own library and the other zone is where books are stored while they are not used and called as archive.

The library presents a gross floor area of 300 m^2, is located in the first plant of the building and presents a porch that reduce solar incidence. This library is occupied by students from 9:00 a.m. to 21:00 p.m. and employ books stored in wooden and metal bookcases as we can see in Figures 1 and 3.

Archive is located in an annex building with three plants and its only access is from stairs located at the end of the library. This archive presents 150 m^2 floor area but do not have opened windows, except a little to supply slight air renovations when the librarian considers it necessary. That archive only presents metal bookcases where books are stored for a future use. No human heat sources were detected in this room.

In both rooms, principal walls are external walls and its construction presents; a layer of concrete, brick, air barrier, insulation, brick and finally painted plaster.

During the year this library present a level of occupation about twenty students but in exam periods all the seats are taken reaching values of one seventy students. This exam period takes, principally, the month of June and July.

HVAC system consists in a water heating system whose exchangers are located in both zones and is only operative during the winter season. There is not a cooling system for the summer season and no exist mechanical air renovation. All the air renovation is that obtained with the windows. It is important to point that, as one of the adaptive model conditions, students can open or close windows when they consider it necessary. This is an ASHRAE Standard 55-2004 condition to employ the adaptive models for predicting indoor thermal comfort temperatures. Furthermore, this standard indicates that occupants may freely adapt their clothing to the indoor and/or outdoor thermal comfort.

Figure 8.2.1.1. Library and Archive.

8.2.2. Students

Students present a mean age of twenty-three years old, a mean clo value of 0.5 and near sedentary physical activities with metabolic rates ranging from 1.0 to 1.3 met. These occupants can freely open and adjust the windows, which open to outdoors.

8.2.3. Sampled Variables

ASHRAE Standard shows that the six primarily factors that must be addressed when define conditions for thermal comfort are: metabolic rate, clothing insulation, air temperature, radiant temperature, air speed and humidity. On the other hand, as was explained, for materials preservation indoor air temperature, dust and relative humidity are the more important parameters.

8.3. METHODS

8.3.1. Standards

To investigate this kind of environment some standards must be considered. In this sense, ASHRAE Handbook Fundamentals 2005 [12] in its chapter forty titled "Codes and standards" remember us the principal standard to consider on HVAC Applications.

The first parameter is the comfort condition. That is defined by the ASHRAE in the ANSI/ASHRAE 55-2004 [13] "Thermal Environmental Conditions for Human Occupancy" which is in close agreement with ISO Standards 7726:1998 "Ergonomics of the thermal environment-Instruments for measuring physical quantities" [14] and the ISO 7730-1994 "Moderate Thermal Environments—Determination of the PMV and PPD Indices and specification of the Conditions for Thermal Comfort" [15]. These standards are principally based on Fanger studies [16].

ASHRAE [12] remember us that there are not established a strict lower humidity limits for thermal comfort and consequently, this standard does not specify a minimum humidity level. On the other hand, this same standard show us that systems designed to control humidity shall be able to maintain a humidity ratio at or below 0.012, which corresponds to a water vapour pressure of 1.910 kPa at standard pressure or a dew point temperature of 6.8 °C.

The second parameter to consider is the materials preservation in museums, libraries and archives that are considered in the ASHRAE Handbook 2003 HVAC Applications [7] in its chapter 21. In our case only books and recently disc and electronic components are of interest.

8.3.2. Temperature and Humidity

To measure temperature the ANSI/ASHRAE 41.1-1986 (RA01) "Standard Method for Temperature Measurement" [17] and the "Temperature Measurement" ASME PTC 19.3-1974 (RA04) [18] were considered. After it, indoor moist air properties were sampled and calculated in accordance with the ANSI/ASHRAE 41.6-1994 (RA01) the "Standard Method for Measurement of Moist Air Properties" and ASHRAE Handbook Fundamentals 2005 [12] in its chapter 6 Psychometrics.

In consequence, building was randomly chosen so that typical every day life of occupants was not disturbed. Indoor temperature and humidity have been measured in occupied zones of the building at the locations where the occupants are known to or are expected to spend their time. In unoccupied rooms, like the archive, the evaluator estimated a good faith of the most significant zone of the room like the centre of the room. Indoor measurements used to be taken with a sampling frequency between five to ten minutes during the months of June and July and outdoor measurements were obtained from the SIAM [19] meteorological stations with a sample frequency of ten minutes.

8.3.3. Adaptive Models

Over the last few years, adaptive models are applied to define the neutral temperature as a function of outdoor, indoor or both temperatures. Some of them present a higher accuracy in certain conditions and, as a result, these principal models were employed for this study. Auliciems and de Dear developed the relations for predicting group neutralities based on mean indoor and outdoor temperatures which were employed by the ASHRAE in Eq. 8.3.3.1. Many other adaptive models have also been proposed [21] and Nicol and Roaf [20].

$$T_c = 17.8 + 0.31 T_o \tag{8.3.3.1}$$

where;

Tc Comfort temperature (°C).

To Arithmetic average of the mean daily minimum and means daily maximum outdoor (dry bulb) temperatures for the month in question (°C).

Before applying these models, we must remember that occupants must be engaged in near sedentary activity (1-1.3 met) and must be able to freely adapt their clothing. Furthermore, neither a heating system nor a mechanical cooling system can be in operation, although non-conditioned mechanical ventilation can be present. Despite this, windows must be the principal way of controlling the thermal conditions.

8.3.4. Thermal Comfort/Indoor Air Quality

In these naturally ventilated building Fanger's thermal comfort indexes present a certain error related with a bad met or clothing indexes estimation. In consequence, it is needed to employ new indexes that let us understand comfort conditions and perception of indoor air quality. Local thermal comfort has been evaluated in terms of the parameter PD, Percentage of Dissatisfied Persons, through the Eq.2.3.5.1 and the Indoor Air quality has been evaluated through the so-called Acceptable Indoor Air Quality parameter, Acc showed in Eq.3.1.2.2. It has been found elsewhere, Fang et al (1988) [27-29], that this parameter is strongly influenced by the temperature and the relative humidity, being linearly related to the air enthalpy.

8.4. RESULTS

Figure 8.4.1 shows outdoor temperature, relative humidity climatic conditions and indoor thermal comfort conditions in accordance with adaptive models. Figures 8.4.2 to 8.4.8 show real indoor sample data.

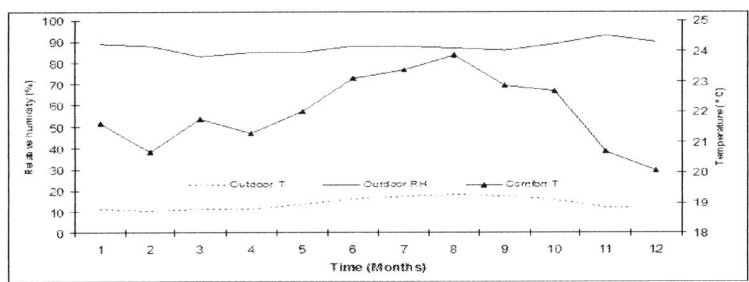

Figure 8.4.1. Outdoor mean temperature and relative humidity and predicted indoor thermal comfort temperature.

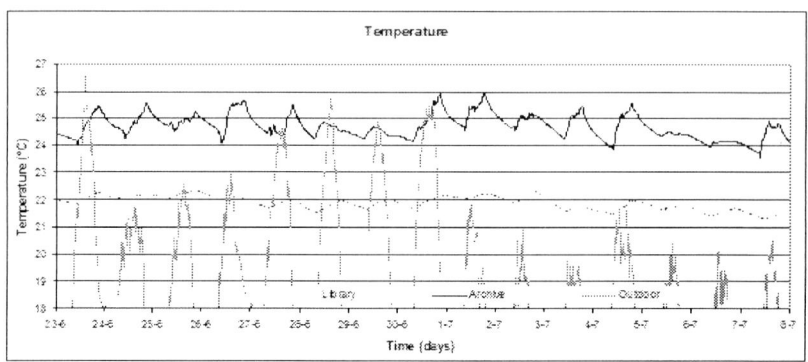

Figure 8.4.2. Indoor weekly temperature conditions.

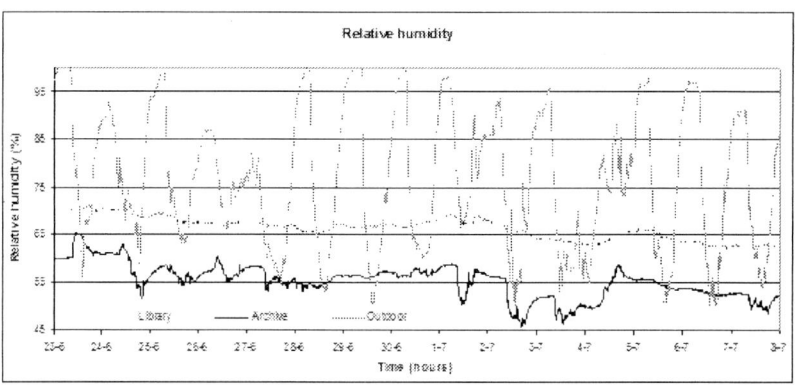

Figure 8.4.3. Indoor weekly relative humidity conditions.

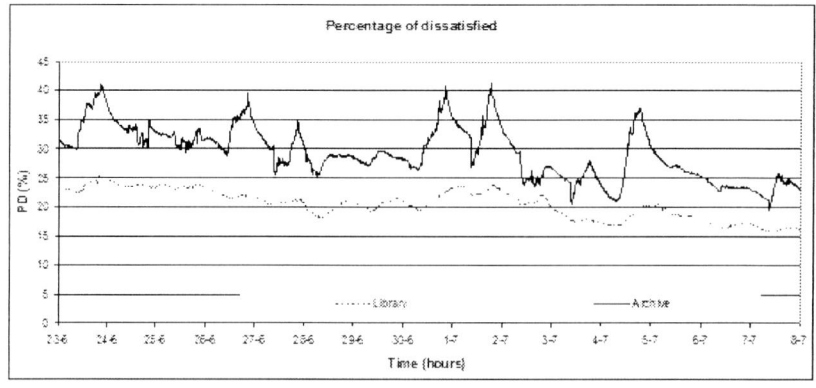

Figure 8.4.4. Percentage of dissatisfied in archive and library.

Indoor Air and Materials Conservancy 141

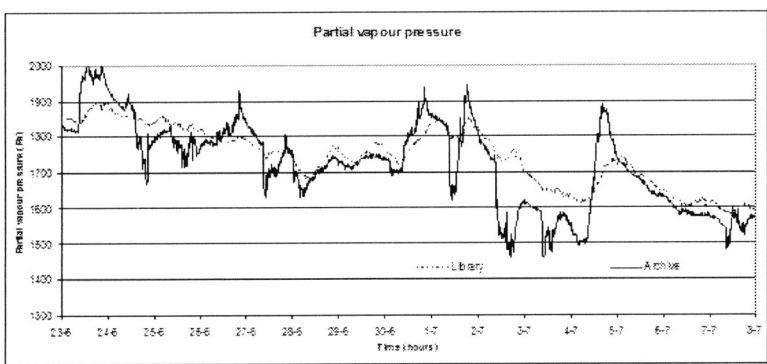

Figure 8.4.5. Indoor partial vapour pressure in archive and library.

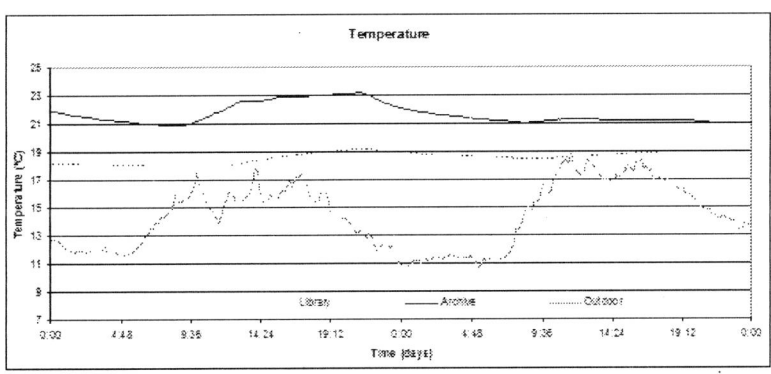

Figure 8.4.6. Hourly indoor/outdoor temperature.

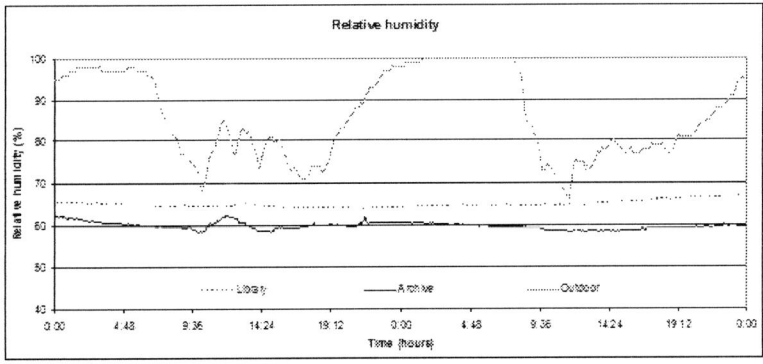

Figure 8.4.7. Hourly indoor/outdoor relative humidity.

8.5. Discussion

A Coruña is located in the northwest of Spain. As a consequence of winds, this Spanish region presents a higher relative humidity during all the year, see Figure 8.4.3. This outdoor relative humidity reaches mean values of 88% during the summer months of June and July and outdoor temperature present a mean value of 17 °C during this same months.

In this chapter a practical case study about indoor conditions in Spanish natural ventilated library was done with the aim to relate indoor ambience with thermal comfort and materials conservancy standards. Our study period, that comprise from June to July, was selected because is when the number of students that occupy the library experiment a clear increment and, in consequence, indoor condition under natural ventilation reach their limit values. In particular, conditions of temperature and relative humidity were expected that exceed standards indications.

Considered environmental levels are based on ASHRAE indications and Environmental Specifications for the Storage of Library and Archival Materials [30] whose recommended environmental ranges for books, paper, and people. For general stacks and branch libraries are of 20-22 °C and a relative humidity between 40-55%. In this standard is explained that rapid fluctuations do the most damage to collections and, in consequence, the goal in this ambiences is to maintain constancy within the specified ranges for temperature, and for humidity, accommodate seasonal fluctuations, but minimize daily and weekly variations.

In this sense, in accordance with the ASHRAE adaptive thermal comfort model of Eq. 8.4.1, indoor comfort conditions were calculated for all the months of the year as we can see in Figure 8.4.1. That figure shows that extreme temperature comfort conditions must be between 20 °C and 24 °C in December and August months respectively. In particular, in the June and July months it must be about 23 °C. When we compare this indoor comfort conditions respect-sampled data we can notice that, indoor temperature in the library do not present a daily swing reaching a mean value of 22 °C during this two months period. In the archive indoor temperature shows a higher mean value of 25 °C during the month of June and July and clear daily temperature change in accordance with the library occupation. This means temperature exceed limits of comfort conditions and materials conservancy will be affected resulting in the paper fragility, especially with humidity change. One of the principal differences between these two ambiences is related with the fact that the library presents a porch that reduces the radiative heat gain. These, and a low air renovation, are the causes of this temperature difference.

If indoor relative humidity is analysed we can observe that, in the library, exist a mean indoor relative humidity of 65% reaching maximum values of 70%. In the archive indoor relative humidity present a lower mean value of 55%, as we can see in Figure 8.4.3. Furthermore, from Figure 8.4.5 we can see that indoor maximum partial vapour pressure limit of the ASHRAE standard was exceeded by the sampled data. This archive presents a partial vapour pressure value of 2000 Pa at the end of June and the beginning of July.

After analyse this sample data we can conclude that, in the library, despite natural ventilation reduce peaks of temperature, due to occupants heat and moisture gains, these gains are send to the archive thought stairs and, in consequence, will affect books preservation. This archive presents a higher temperature mean value because there is not a higher external air renovation and exist a solar radiant gain that affect its walls but, as was explained, it is not the only heat and moisture source. In consequence, library sends its hourly gains through the stairs during the occupied period, 9:00 to 21:00, as we can see in Figures 8.4.6 and 8.4.7. This is proved if we compare indoor ambiences during the weekend, when this difference is reduced and exist a clear tendency to converge.

Despite the fact that naturally ventilation zones must not be analysed respect the PMV and PPD indexes, if we compare our indoor conditions respect ASHRAE/ISO thermal comfort limits, we can conclude that, indoor conditions are in the upper limit of the comfort zone. If now we employ the percentage of dissatisfied index we can conclude that comfort conditions exceed its limit value of 15% of dissatisfied in both environments and that the archive present the higher mean value reaching to 40% in some days, see Figure 8.4.4. Despite this, occupants present a certain tolerance to this high relative humidity of the library as was expected for natural ventilated building and contrasted with librarian's indications and experiences.

In the archive an approach to materials preservation is done. In this sense, indoor air relative humidity present a mean value of 58% under a mean temperature of 25 °C. Despite the fact that this relative humidity is not so high than in the library, its excessive temperature let mould growth. This fungal growth was detected in wall constructions and books damages and related with a high relative humidity for a long period of time, a lower air speed and a clear presence of dust.

Future works about design corrections to improve indoor ambiences under naturally ventilated buildings must be done. In particular, it is interesting the possibility of night cooling by mechanical ventilation to reduce peaks of discomfort conditions during the occupied period and a lower temperature of 19ºC at first hours of the morning. Other possible solution are the passive climate control [31] by permeable coverings or passive solar design as low energy cost solution that let us control indoor relative humidity and temperature, especially during the unoccupied period.

Conclusions

In this chapter a practical case study about indoor conditions in a natural ventilated library was done with the aim to relate indoor ambience with thermal comfort standards and materials conservancy.

Results showed that A Coruña presents an higher relative humidity during all the year an that this relative humidity affect indoor comfort conditions and materials preservation. In particular, during the exams period library is fully occupied and temperature and relative humidity gains affect the archive condition increasing mould risk. Possible solutions for the archive problem are mechanical ventilation and permeable coverings to reduce indoor temperature and relative humidity at first hours of occupation or solar design corrections to reduce solar heat gain.

References

[1] Krausse B. Cook M. Lomas K. Environmental performance of a naturally ventilated city centre library.2007. *Energy and Buildings.* 39, 792-801.

[2] Lomas K.J. Architectural design of an advanced naturally ventilated building form.2007. *Energy and Buildings.*29, 166-181.

[3] Finn D.P., Connolly D., Kenny P. Sensitivity analysis of a maritime located night ventilated library building. 2007. *Solar Energy.* 81, 697-710.

[4] Chow T.T., Clarke J.A. Dunn A. Primitive parts: an approach to air-conditioning component modelling.1997. *Energy and Buildings.* 26, 165-173.

[5] Kalagasidis A.S., Weitzmann P., Nielsen T.R., Peuhkuri R., Hagentoft C., Rode C. The International Building Physics Toolbox in Simulink.2007. *Energy and Buildings.* 39, 665-674.

[6] Pavlogeorgatos G. Environmental parameters in museums. 2003. Building and Environments. 38, 1457-1462.

[7] ASHRAE HandbookHVAC Applications, 2003.

[8] Canhoto O., Pinzari F., Fanelli C., Magan N. Application of electronic nose technology for the detection of fangal contamination in library paper. 2004. International *Biodeterioration and Biodegardation.* 54, 3030-309.

[9] Ipert S., Dupont A.L., Lavédrine B., Bégin P., Rousset E., Cheradame H. Mass deacidification of papers and books. IV- A study of papers treated with aminoalkylalkoxysilanes and their resistance to ageing. 2006. *Polymer degradation and stability.* 91, 3448-3455.

[10] Schieweck A., Lohrengel B., Siwinski N., Genning C., Salthammer Organic and inorganic pollutants in storage rooms of the Lower Saxony State Museum Hanover, 2005. Germany. *Atmospheric Environment.* 39, 6098-6108.

[11] UCSD Presenrvation. http://orpheus.ucsd.edu/preservation. [Accessed december 2010].

[12] ASHRAE Handbook Fundamentals, 2005.

[13] ANSI/ASHRAE 55:2004 Thermal Environmental Conditions for Human Occupancy.

[14] ISO 7726:1998 ergonomics of the thermal environment- Instruments for measuring physical quantities.

[15] ISO 7730:1994 Moderate thermal environments- Determination of the PMV and PPD indices and specification of the conditions for thermal comfort.

[16] Fanger P.O. Thermal comfort. Analysis and applications in environmental engineering.1970. McGrawHill. ISBN:0-07-019915-9

[17] ANSI/ASHRAE 41.1-1986 (RA01) "Standard Method for Temperature Measurement"

[18] ASME PTC 19.3-1974 (RA04) Temperature Measurement.

[19] Environmental Information system of Galicia (SIAM). http://www.Siam-cma.org.

[20] Nicol F. Roaf S. Pioneering new indoor temperature standard: the Pakistan project. Energy and Buildings. 1996. 23, 169-74.

[21] Humphreys MA. Comfortable indoor temperatures related to the outdoor air temperature. 1976. *Building Service Engineer.* 44. 5-27.

[22] Simonson C.J., Salonvaara M., Ojalen T. The effect of structures on indoor humidity-possibility to improve comfort and perceived air quality. 2002. *Indoor Air*. 12. 243-251.

[23] Simonson CJ., Salonvaara M.,. Ojalen T. Improving indoor climate and comfort with wooden structures. 2001. Espoo. Technical Research Centre of Finland, VTT Publications 431.200p.+ app 91 p, 2001.

[24] Orosa JA, Baaliña A. Passive climate control in Spanish office buildings for long periods of time. Building and Environment (2008). doi:10.1016/j.buildenv.2007.12.001

[25] Toftum J., Jorgensen A.S., Fanger P.O. Upper limits for indoor air humidity to avoid uncomfortably humid skin. 1998. *Energy and buildings*. 281-13.

[26] Toftum J., Jorgensen A.S., Fanger P.O. Upper limits of air humidity for preventing warm respiratory discomfort.1998. *Energy and buildings*. 28.15-23.

[27] Fang L., Clausen G., Fanger PO. Impact of Temperature and Humidity on the Perception of Indoor Air Quality.1998. *Indoor Air*. 8. 80-90.

[28] Fang L., Clausen G., Fanger PO. Impact of Temperature and Humidity on Perception of Indoor Air Quality During Immediate and Longer Whole-Body Exposures.1998. *Indoor Air*. 8.276-284.

[29] Fang L., Wargocki P., Witterseh T., Clausen G., Fanger P. O. Field study on the impact of temperature, humidity and ventilation on perceived air quality.1999. The 8th International Conference on Indoor Air Quality and Climate. Edinburgh, Scotland. 1999. 2.107. 8–13,.

[30] Solinet. http://www.solinet.net/emplibfile. [Accessed December 2010]

[31] Calderaro V., Agnoli S. Passive heating and cooling strategies in an approaches of retrofit in Rome.2007. *Energy and Buildings*. 39, 875-885.

Chapter 9

INDOOR AIR AND WORK RISK: A PRACTICAL CASE STUDY OF INDUSTRIAL ENVIRONMENT

ABSTRACT

Ships are a clear example of thermal indoor environment with characteristics changing in short intervals of time. In the engine room the conditions used to be extreme as well.

To analyse this industrial ambience, it was carried out a monitoring of air temperature and relative humidity in several locations of a merchant ship.

Subsequently, it has been determined from those indoor temperature and relative humidity data, the corresponding parameters of thermal comfort (predicted mean vote, PMV; predicted percentage of dissatisfied, PPD and acceptability, Acc) and heat stress of sever exposures in the engine room.

9.1. INTRODUCTION

The thermal parameters of indoor environments need a suitable study, knowledge and processing in relation with the industrial safety, because high or low air temperatures and not controlled heat sources, may induce lower productivity rates, higher accidents rates and health hazards.

A clear example of thermal indoor environment with extreme characteristics changing in short intervals of time is a ship. In this chapter, we have carried out the sampling of air temperature and relative humidity in a merchant ship.

In the indoor environment of a ship, it is the engine room that is important having in mind its special characteristics related with safety at work. Temperature and relative humidity data from the engine room and other locations have been analysed to obtain comfort indexes by means of new models applied to this environment. From this analysis of real conditions, it has been possible to define limit time that a person can work without heat stress in accordance with Spanish standards NTP.

9.2. METHODS

By means of Gemini® data loggers, a monitoring of air temperature and relative humidity has been carried out in a merchant vessel during the winter season of 2001. The engine room together with the control room, the dinning room for officers and the bridge has been the sampling locations. At the same time, outdoor data have been also obtained for comparing purposes. More than 11,000 measurements have been collected.

As previously stated, we have paid special attention to the engine room environment. Thus, we have collected data both from the control engine room and the proper engine room.

To obtain measurements of work environments, data-loggers were located to a height near the centre of gravity of workers when they remain in the usual position of work. In those places where this position was not possible to fix, the sampling points were moved away from heat sources such as walls or air conditioning equipments at least 0.6 m, for avoiding interferences. This is the case of the engine room where several work places, those were anticipated hotter and colder and in the centre of place, have been assessed according to recommendations of INNOVA [4].

9.3. ANALYSIS OF MONITORED VARIABLES

Because ASHRAE standards [2] only discuss about the different comfort zones, the European standard ISO 7730 [5] has been analysed in relation to the required conditions of indoor environments. Table 9.3.1 lists these conditions for different seasons and all practical applications.

Figures 6 and 7. Control engine room and engine room.

As opposed to the real conditions in other locations, it has been obtained an average temperature in the control engine room of 19.76 °C. This temperature is too low compared to that in the engine room, so this fact can cause both a thermal shock for workers and high-energy consumption for air conditioning.

Average relative humidity in the bridge and dinning room has reached 50%, very close to the maximum value recommended in standards. Whereas in the control engine room average relative humidity was 40%, in the engine room the value was 25%, in clear opposition to the rest of locations and under the minimum value of 30% recommended.

Table 9.3.1. Reference values for indoor conditions

Applications	Summer		Winter	
	Temperature (°C)	Relative humidity (%)	Temperature (°C)	Relative humidity (%)
All	>23	30-65	18-22	30-65

Given the influence of outdoor conditions on indoor environment, we have collected outdoor data during the monitoring period from December 2001 to February 2002. The results obtained were an average temperature of 23°C, with maximum values of 27 °C and an average relative humidity of 60%. The weather was predominantly sunny and not much cloudy.

The assessment of indoor environments is based in the study of comfort indexes, however the variables involved in the definition of such indexes must remain in the range provided by standards. For this reason, we have carried out a statistical analysis of temperature, relative humidity and enthalpy data. Results showed that the average temperature of the air both in the bridge and dinning room has remained about 22 °C. In the engine room the average was 32.5 °C, with peaks of 38.5 °C. These results are out of the allowed values for hot environments and can produce different health disorders, see Figure 9.3.1.

Temperature	Biologic response	Heat disorders
20 °C	Comfortable	Full capacity
↓	Discomfort Irritability Concentration difficulties Decrease in intellectual capacity	Psychical disorders
	Increase in work mistakes Decrease in handiness More accidents	Psychical and physiological disorders
	Decrease in efficiency of heavy works Disturbance of metabolism Cardiac-circulatory system overload Heavy fatigue	Physiological disorders
35-40°C	Maximum temperature bearable	

Figure 1. Influence of temperature on heat illness.

9.4. STUDY OF COMFORT CONDITIONS

To keep thermal comfort and avoid disorders two conditions must be fulfilled. The first is that the combination of skin and deep body core temperatures leads to a neutral feeling of comfort. The second involves the energy balance between the body and the environment. In this sense, the total metabolic heat produced by the body should be equal to the heat loss from the body.

The comfort equation developed by P. O. Fanger [3] relates physical parameters that can be measured with the neutral thermal feeling experimented by a "typical" person, see Eq. 2.2.2.2.

Through measurement of physical parameters, the comfort equation provides an operative tool whereupon it can be assessed under what conditions the thermal comfort in an occupied space is achievable. The thermal comfort can be quantified through indexes defined in standard ISO 7730 [3]. The Predicted Mean Vote (PMV) is derived from the heat balance before mentioned and provides an indication of the thermal sensation by means of a scale of 7 points, from -3 (cold sensation) to +3 (hot sensation), where 0 means a neutral thermal sensation. Another comfort index is the Predicted Percentage of Dissatisfied (PPD) that provides information on thermal sensation by predicting the percentage of people likely to feel too hot or too cold in a given environment. PMV values of -3, -2, +2 and +3, means thermal discomfort in the PPD index. Both indexes are influenced by physical activity and clothing. The physical activity is quantified through the metabolic rate. On the other hand, clothing acts as an insulation reducing the heat loss from the body.

In this study, we have used the PMV models developed by the Institute for Environmental Research Kansas State University [1] for the assessment of the thermal comfort conditions in indoor environments. The expression of the model was showed in 2.2.3.1.3 and now in the Eq. 9.4.2.

$$PMV = at + bP_v - c \qquad (9.4.2)$$

where t is the temperature and Pv is the vapour partial pressure. Right constants a, b and c must be used to take into account sex and time of exposure to the indoor environment. Such constants have been adapted to the existing conditions in the studied environments by means of a thermal comfort datalogger 1221 from Innova. Values of 1.2 Met and 1 Clo were assumed for calculations.

PPD has been also studied for the same environment. The index has been defined by means the Eq. 2.2.2.6, taking into account the PMV values previously obtained.

Temperature and relative humidity values collected in the ship have been introduced in the models just detailed and corresponding PMV and PPD values has been calculated for each indoor location.

In order to make comparisons, the thermal acceptability (Acc) has been also calculated. This new index, introduced by Fanger [3], has been used by Simonson [8] to assess indoor environments as a result of their adaptation to any kind of thermal conditions but with loss of accuracy. The index is related with enthalpy through the Eq. 3.1.2.2.

Once averages and standard deviations have been assessed, it has been quantified what values have been broken the standards according with ASHRAE [2]. Such specifications set a PMV range that goes from -0.5 to +0.5 as adequate. This interval is equivalent to a PPD lower than 10 %.

Bridge and dining room

Cumulative PPD in the bridge and dining room were close to the neutral thermal conditions. The last one is the location more comfortable, followed by the bridge; since 90% and 78% of collected data are included in the 10% of PPD, according with the ASHRAE standard [2]. These results mean that it is not possible the energy optimisation of indoor environment, because outdoor and indoor air temperatures maintain similar values. Nevertheless, in these environments relative humidity usually exceeds 55% established by ASHRAE and ISO standards [2,3]. For this reason, it would be advisable the use of dehumidifiers for protecting both occupiers and electrical devices of the bridge.

Engine room

The most extreme conditions have been found in the engine room. PMV and PPD values of 2.15 and 76.61%, has been obtained. These extreme values together with the fact that nearly all calculated PPD data exceed 10% show that in this location the thermal sensation is very hot.

From Simonson's studies [8] can be deduced that acceptability is better with a lower enthalpy. He has established the value of 50 kJ/kg as the upper limit for enthalpy. Above this value the air perception is unbearable, independently of the indoor air quality. In our case, the enthalpy value in the engine room exceeds 52 kJ/kg and thus the calculated acceptability has been -0.05.

Values of relative humidity are low because high outdoor temperature. These conditions may cause hyperthermia, vasodilatation, sweat glands activation, increase of peripheral circulation and electrolytic changes of sweat by loss of salt content. As a possible solution to this problem, an increase in ventilation rates may achieve some decrease in temperature.

Control engine room

In this location the PMV is close to 0.5, the optimum condition for energy saving, the average PPD is 13% and 40% of calculated PPD data are within 10% fixed in standards. These results lead to a limit condition in the control engine room. Besides, the average temperature of 19.76 °C and the average relative humidity of 24.9% (lower than the minimum reference value of 30 % from ISO 7730 [3]) are too low which means a high-energy consumption in air conditioning.

To get an energy saving, the temperature should be higher than 20 °C as in accordance to the air acceptability criteria, to maintain the same PPD, the enthalpy must be the same as well. For this reason, a temperature increase of 1 °C leads a relative humidity decrease of 5% and the lower limit of 30% already mentioned would not be fulfilled. All these facts lead to set the existing temperature conditions as optimum for the existing relative humidity.

A possible improvement may be to increase the air renovations with outdoors, to cause an increase of both relative humidity and temperature towards values more suitable.

Another solution to correct the low relative humidity may be to avoid the possible dehumidification of conditioning air in the chiller by means air conditioning unit to be replaced by another one with higher surface area of heat exchange that causes a higher external surface temperature for the same heat rate.

9.5. WORK RISK ANALYSIS

As fast as indoor temperature increases the first psychical disorders appear such as loss or difficulties of concentration. Finally, physiological disorders as heart and circulatory system overload could yield. According to the indications of NTP 18 (Heat stress evaluation of sever exposures) [6, 8, 9] and NTP 350 (Heat stress evaluation required sweating index) [7] standards, we got the corresponding graphics in Figures 9.5.1 and 9.5.2. They are based in human body thermal balance and they showed the maximum time that a worker could remain in severe exposures as engine rooms. The minimum time that the same worker must be at the control room to lose the accumulated heat was also calculated.

This study is referred to a standard worker with 70 kg in weight and light clothes. The Figures show that the exposition must stop when the internal temperature increases 1 °C, because the maximum evaporation is lower than the required for the thermal balance.

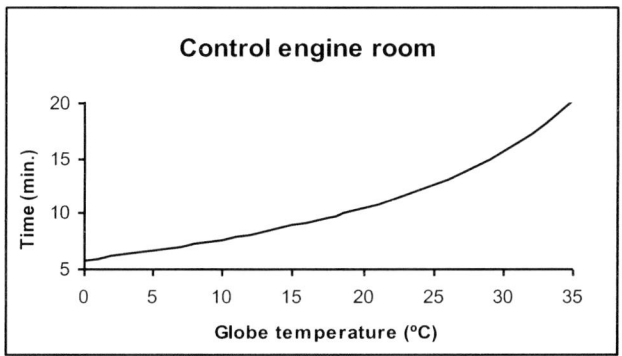

Figure 9.5.1. Minimum time that must be elapsed in the control engine room.

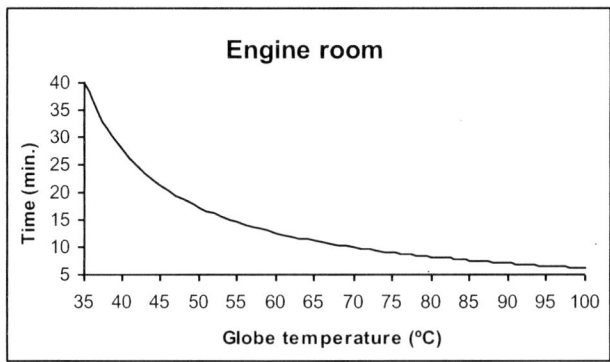

Figure 9.5.2. Maximum time that must be elapsed in the engine room.

These figures express the relationship between time and globe temperature because the ship ambient changes during the voyage. Using the obtained graphics for this engine room we can affirm that the worker must be in the engine room for 17 minutes and must have a rest at the control room for at least 10 minutes in order to get the suitable heat release.

CONCLUSIONS

Ships show very different thermal environments that must be studied with greater depth. Options of energy saving or thermal comfort improving in the bridge and the dinning room are very limited. The engine room shows air conditions out of any recommendations from standards, so it is suggested an increase in ventilation for taking preventive measures against work risks.

The control engine room shows limit conditions of thermal comfort with temperature values too low that lead to an excessive energy consumption. As the outdoor air conditions are suitable, an increase of renovations with outdoor air can be proposed.

Taking into account our results it would be necessary to take the following work risk-preventing measures:

- Drinking water. Sources of drinking water must be available close to work locations and workers must be informed about the necessity of drinking frequently.
- Acclimatization. Workers starting new or going back to work require an exposure time for achieving acclimatization.
- Metabolic heat. Adjusting length and frequency of breaks and work periods, and work rates may be reduced the metabolic heat release. If it is possible works must be scheduled in time of less heat.
  Work periods into engine room must not be higher than twenty seven minutes and, after it, worker must be about ten minutes in the control room.
- Workers must be kept under constant watch by a trained colleague for detecting any symptom of heat strain.

As we can deduce form this study about real industrial ambiences, ships present great different indoor ambiences that change in short periods of time and present a new research area for the next years.

REFERENCES

[1] ASHRAE. HVAC Fundamentals.1985. ASHRAE. Atlanta.
[2] ASHRAE STANDARD. Proposed Revision to an American National Standard Thermal Environmental Conditions for Human Occupancy. 2001. FIRST PUBLIC REVIEW DRAFT.
[3] Fanger, P.O. Thermal comfort analysis and applications in environmental engineering. 1970. USA: McGraw-Hill.
[4] INNOVA Air Tech Instruments A/S. Thermal Comfort. 1997.. Denmark. Available from: http://www.innova.dk/books/ thermal/ [Accesed 1 December 2010]
[5] ISO 7730. Moderate thermal environments. Determination of the PMV and PPD indexes and specification of the conditions for thermal comfort.
[6] NTP 18 (Heat stress evaluation of sever exposures). Normativa Técnica de Prevención (NTP). Ministerio de Industria. http://www.insht.es/
[7] NTP 350 (Heat stress evaluation required sweating index). Normativa Técnica de Prevención (NTP). Ministerio de Industria. http://www.insht.es/.
[8] Simonson et al.. Improving indoor climate and comfort with wooden structures. 2001. Technical research centre of Finland. Espoo.
[9] ISO 7793. Ergonomía del ambiente térmico. Determinación analítica e interpretación del estrés térmico mediante el cálculo de la sobrecarga térmica estimada.
[10] ISO 11064-6:2005 Ergonomic design of control centres- Part 6: Environmental requirements for control centres. 2005.

THANKFULNESS

I want to express my gratitude to LumaSense Technologies, Tiny Tag, PCE Group, Casella Ltd., INEGA, DEMEGI, Xunta de Galicia (Europan Social Fund (ESF) and the University of A Coruña and all individuals and institutions that collaborated during the writing of this book.

A Coruña, 2009

The author, José Antonio Orosa García is graduated in Marine Engineering and Naval Architecture at the University of A Coruña. He is the prize-winner of the studies of Master in Marine Engineer. It's Ph.D. about modelling thermal comfort and air renovation for energy saving in buildings has proved the possibilities of energy saving with permeable coverings.

Nowadays, he is professor of the Department of Energy of the University of A Coruña. During last years, as ASHRAE member, he has participated in the International Energy Agency Annex 41 "Whole Building Heat, Air, Moisture Response" and collaborates with the INEGA (Energetic Institute of Galician) and IDEMEC (Faculty of Engineering of the University of Porto) in HVAC research and transfer of knowledgement to companies.

INDEX

A

access, 83, 134, 135
accessibility, 41
acclimatization, 155
accounting, 92, 114
activity level, 3, 30
adaptation, 5, 8, 12, 17, 38, 42, 152
adhesives, 49
adjustment, 8
adsorption, 83
adults, 54
adverse effects, 52
aerosols, 47, 50, 51, 53
aesthetic, 133
age, 4, 29, 136
air pollutants, 6, 46
air quality, 1, 3, 5, 6, 13, 16, 21, 37, 41, 42, 45, 46, 53, 54, 56, 57, 58, 62, 63, 64, 65, 76, 79, 90, 91, 93, 95, 96, 104, 111, 128, 132, 139, 146, 152
air temperature, 4, 18, 20, 21, 31, 32, 33, 34, 38, 42, 68, 72, 73, 74, 93, 104, 115, 124, 125, 132, 137, 145, 147, 148, 152
allergens, 10, 11, 13, 14, 47
allergic asthma, 52
allergic rhinitis, 52
allergy, 9, 13, 14
aluminium, 74

ambient air, 32, 38, 49, 79, 89
ambient air temperature, 32
American Conference of Governmental Industrial Hygienists, 53
American Society of Heating Refrigerating and Air Conditioning Engineers), 52
amplitude, 125
arithmetic, 85
arthropods, 52
asbestos, 48, 49
ASHRAE Standard, 4, 5, 12, 40, 41, 59, 62, 77, 80, 96, 115, 131, 132, 135, 137
assessment, 5, 9, 41, 63, 79, 80, 150, 151
assets, 134
asthma, 9, 10, 11, 13
asymmetry, 70, 78
athletes, 47
atmosphere, 11, 19, 24, 52, 59, 134
audit, 93, 95
authorities, 53, 101, 134
authority, 103
avoidance, 10

B

bacteria, 63
bacterium, 52
banks, xiii
base, 11

bedding, 10
benchmarking, 92
benchmarks, 53
blood, 48
boreholes, 92
brain, 16
Brazil, 84
breathing, 59, 101
breeding, 52
Building Energy Software Tools Directory, 81, 97

C

CAD, 93
calibration, 92
cancer, 50
carbon, 46, 47, 48, 59, 101, 103, 111
carbon dioxide, 46, 47, 48, 59, 101, 103, 111
carbon monoxide, 46, 48
Carbon Monoxide (CO), 78
case studies, 91
case study, 101, 103
causality, 9
cellulose, 102, 134
chemical, 10, 46, 47, 48, 51, 54, 132, 134
chemical reactions, 132
chemicals, 46, 48, 49, 53
children, 9, 13, 46, 51
chlorinated hydrocarbons, 50
cigarette smoke, 50
circulation, 32, 78, 108, 153
class period, 116
classification, 17
classroom, 89, 116, 118, 124, 125
clean air, 5, 6, 8, 38, 52, 55, 57
cleaning, 47, 49, 63, 109
climate, 11, 16, 81, 82, 91, 93, 95, 101, 113, 114, 127, 128, 131, 132, 134, 144, 146, 156
climates, 115

clothing, 8, 18, 20, 26, 30, 31, 38, 40, 58, 132, 136, 137, 139, 151
CO_2, 47, 48, 53, 54, 76, 77, 78, 91, 103, 104, 108, 110
coatings, 22, 45, 49
cogeneration, 91
colonisation, 133
combustion, 48
commercial, 45, 82, 83, 91, 92, 93, 94, 95, 96, 97
Committee for Bioaerosoles, 53
communities, 85
community, 81
compensation, 75
compilation, xiii
complement, 67
complexity, 133
compliance, 91, 92, 93, 94, 95
composition, 133
compounds, 50
comprehension, 1
computational fluid dynamics, 95
computer, 9, 67, 81, 82, 84
condensation, 10, 70, 96, 109
conditioning, xiii, 15, 16, 21, 32, 45, 49, 50, 52, 59, 63, 91, 93, 95, 96, 113, 114, 125, 132, 144, 148, 149, 153
conduction, 126
conductivity, 36, 86, 87, 89
conservation, xiii
construction, 10, 12, 45, 49, 74, 82, 84, 89, 118, 119, 126, 135
consumption, 15, 81, 113, 114, 115
contaminant, 95, 96
contamination, 49, 55, 63, 133, 145
cooling, 8, 31, 32, 33, 40, 47, 51, 57, 70, 73, 81, 82, 83, 84, 89, 91, 92, 93, 94, 114, 115, 118, 132, 135, 139, 144, 146
copper, 74
correlation, ix, xi, 9, 11, 124
correlations, 29
corrosion, 50, 134
cosmetics, 49

cost, 82, 83, 85, 91, 92, 93, 94, 144
cotton, 73
cough, 45, 52
covering, 82, 123, 124, 125
cracks, 50
critical value, 133
crystals, 11
culture, 104
cycles, 134

D

damping, 64
danger, 132
database, 82
decay, 50, 60, 61, 64, 115, 125, 133
decoration, 49
defence, 135
degradation, 49, 133, 145
dehydration, 41
Denmark, 42, 156
Department of Commerce, 128
Department of Energy, ix, xi, 12, 83, 90, 97, 115, 157
deposition, 134
deposits, 50
depth, 12, 155
desorption, 83
detection, 63, 75, 78, 145
developed countries, 21
deviation, 29, 116
dew, 71, 137
diffusion, 11, 87
discomfort, 32, 40, 43, 45, 46, 69, 70, 128, 144, 146, 151
diseases, xiii, 51, 52, 63
dissatisfaction, 8, 16, 32, 33, 34, 35, 36, 37, 38, 53, 55, 56
distilled water, 73
distribution, 60, 78, 95
dizziness, 45
dosage, 61

draft, 32
draught, 33
drinking water, 155
drying, 134

E

education, 93, 95, 96
emission, 46, 48, 60, 61, 62, 74, 133
endotoxins, 52
endurance, 134
energy, ix, xi, xiii, 5, 8, 12, 13, 15, 17, 21, 23, 24, 32, 57, 59, 60, 65, 76, 81, 82, 83, 84, 86, 87, 89, 91, 92, 93, 94, 95, 96, 97, 99, 113, 114, 115, 116, 118, 126, 127, 128, 131, 132, 144, 149, 151, 152, 153, 155, 157
energy consumption, 13, 21, 32, 59, 60, 82, 91, 113, 114, 149, 153, 155
energy efficiency, 92, 95, 128
energy transfer, 113, 115
engineering, 12, 145, 156
environment, xiii, 4, 5, 8, 10, 11, 13, 14, 15, 16, 17, 18, 20, 21, 25, 26, 27, 28, 29, 32, 37, 38, 39, 40, 41, 42, 45, 46, 49, 51, 53, 57, 59, 61, 62, 64, 72, 74, 79, 80, 81, 85, 94, 96, 103, 116, 127, 133, 137, 145, 147, 148, 150, 151, 152
environmental characteristics, 11
environmental conditions, 15, 39, 40, 42, 134
environmental control, 132
environmental factors, 133
Environmental Protection Agency (EPA), 53, 94
epidemiologic studies, 9
equilibrium, 39, 74
equipment, 51, 61, 67, 83, 91, 93, 94, 96
ergonomics, 41, 145
Europe, 55, 63
European Collaborative Action, 46
evaporation, 32, 38, 39, 73, 87, 154

evidence, 9
evolution, 16, 27, 65, 124
execution, 49
exercise, 47
expertise, 82
exposure, 5, 6, 8, 9, 10, 11, 29, 38, 39, 45, 50, 52, 53, 54, 59, 64, 151, 155
extraction, 59

F

faith, 78, 138
fever, 9, 52
field tests, 9
films, 133
filters, 10, 75
filtration, 52, 62
Finland, 42, 64, 111, 128, 146, 156
flaws, 59
flexibility, 85
fluctuations, 32, 64, 142
fluid, 96
footwear, 36
formaldehyde, 49
foundations, 50
fragility, 134, 142
fragments, 10
freezing, 11
fungi, 9, 10, 11, 13, 14, 51, 52, 57, 63, 67, 76, 101, 109, 110

G

genus, 9
geometry, 81
Germany, 145
germination, 133
glue, 49
gravity, 148
growth, 11, 50, 76, 133, 143
growth rate, 133
guidelines, 52, 63

H

hazards, 63, 147
headache, 45
health, ix, xi, 9, 13, 21, 41, 45, 50, 51, 59, 62, 101, 104, 147, 150
health care, 41
health condition, 101
health effects, 50
health problems, 51
health risks, 9
heart rate, 41
heat capacity, 36, 118
heat illness, 150
heat loss, 19, 21, 24, 25, 26, 27, 71, 72, 73, 91, 94, 118, 119, 151
heat release, 154, 155
heat transfer, 19, 26, 93, 95, 126
Heat, Air and Moisture (HAM), 86
height, 35, 148
hemisphere, 34
history, 9, 13
homes, xiii, 10, 13, 45, 95
host, 84
House, 6, 14
house dust, 10, 11
housing, xiii
human, 3, 15, 18, 21, 33, 38, 40, 42, 46, 47, 48, 51, 55, 56, 59, 62, 72, 135, 153
human body, 3, 15, 18, 21, 33, 38, 47, 72, 153
human perception, 59
humidity, xi, 4, 5, 6, 7, 8, 9, 10, 11, 12, 14, 15, 18, 20, 21, 22, 23, 27, 29, 37, 38, 42, 43, 45, 57, 58, 62, 64, 70, 72, 75, 76, 78, 81, 82, 84, 95, 96, 101, 102, 103, 104, 105, 106, 108, 109, 110, 111, 113, 114, 115, 118, 123, 124, 125, 128, 131, 132, 133, 134, 135, 137, 138, 139, 140, 141, 142, 143, 144, 146, 147, 148, 149, 150, 152, 153
HVAC systems, 10, 16, 85, 93, 96, 126, 127

hybrid, 93, 94
hygiene, 10, 63, 109
hypersensitivity, 52
hyperthermia, 153

I

ideal, 22, 23, 134
identification, 12, 46
illumination, 115
immune response, 51
immune system, 51
improvements, 115
incidence, 102, 108, 135
incomplete combustion, 48
independent variable, 119
individual character, 126
individual characteristics, 126
individuals, 6, 8, 16, 28, 32, 34, 38, 104, 157
Indoor Air Quality (IAQ), 78
indoor cockroach allergen exposure, 11
induction, 45
industrial environments, 45, 50, 99
industry, 38, 48
inertia, xi, 113, 114, 115, 116, 118, 124, 125, 126, 127
infancy, 10
insects, 50
institutions, 134, 157
insulation, 4, 18, 20, 36, 40, 49, 109, 116, 125, 126, 135, 137, 151
integration, 86
interface, 83, 91
interference, 68, 74, 75
internal environment, 52
International Council of Building Research, 52
International Energy Agency, xiii, 12, 115, 127, 128, 133, 157
interoperability, 93
interpersonal relations, 46
interpersonal relationships, 46

intervention, 10
investment, 92
IRC, 12
iron, 50
irradiation, 114, 118
isolation, 29, 48
issues, 7, 61, 115

J

Japan, 55

L

laboratory tests, 57
lead, 9, 50, 125, 126, 133, 153, 155
leakage, 119
learning, xiii, 91
LED, 70
legs, 11, 32
life cycle, 83, 92, 93
light, 38, 46, 70, 75, 131, 132, 154
light emitting diode, 70
Luo, 13
Luxemburg, 63
lying, 72

M

magazines, 134
magnitude, 16, 38, 39, 45, 133
majority, 9, 53, 133
malaise, 52
man, 5, 38, 80
management, 93, 95
manganese, 50
manual procedures, 27
mass, 23, 25, 87, 88, 92, 94, 113, 114, 115, 116, 134
materials, xiii, 8, 10, 11, 45, 46, 48, 49, 50, 63, 82, 96, 99, 131, 132, 133, 134, 137, 142, 143, 144

matter, iv
measles, 51
measurement, 41, 62, 68, 72, 73, 75, 78, 79, 103, 110, 151
measurements, 12, 34, 40, 70, 77, 78, 79, 83, 102, 108, 138, 148
mechanical properties, 134
mechanical ventilation, xi, 28, 31, 93, 132, 139, 144
media, 48
medicine, 64
mercury, 49
Mercury, 91
Metabolic, 155
metabolism, 3, 17, 24, 39
metabolites, 10, 51
metals, 134
methodology, 65, 67, 79, 102
microbiota, 102
microorganism, 51
microorganisms, 10, 51, 53, 109, 133
modelling, 84, 85, 91, 92, 93, 94, 95, 144, 157
models, xiii, 9, 31, 34, 59, 65, 84, 85, 86, 89, 114, 135, 138, 139, 148, 151, 152
modifications, 85, 116
modules, 75
moisture, 10, 26, 37, 82, 83, 84, 85, 86, 87, 88, 89, 91, 93, 96, 114, 115, 124, 133, 143
moisture capacity, 89
moisture content, 88, 133
mold, 52, 67
mood swings, 46
mucous membrane, 8, 38, 54
mucous membranes, 8, 54
mucus, 52
multiplier, 51
multivariate analysis, 11
museums, xiii, 12, 133, 134, 137, 145
music, 133
mycotoxins, 52

N

National Bureau of Standards, 128
National Institute for Occupational Safety and Health, 58
nausea, 45
NCS, 102
neural network, 133
neutral, 3, 4, 12, 16, 18, 24, 28, 31, 33, 35, 46, 54, 138, 151, 152
nitrogen, 48
Nitrous Oxide (N2O), 78
noble gases, 50
nodes, 25
non-smokers, 46
nuisance, 45

O

oil, 48, 49
opportunities, 93
optimization, xiii, 91, 92, 93
ordinary differential equation (ODE), 85
organic compounds, 48, 49
organism, 51
oxygen, 48
ozone, 50

P

paints, 49
Pakistan, 42, 145
parallel, 125
participants, 4
PCR, 14
Percentage of Dissatisfied Persons, 103, 118, 139
permeability, 87, 89
phenol, 49
phosphate, 50
physical activity, 19, 58, 151

physical properties, 89
physics, 85, 86, 89, 97, 128
Physiological, 43
plants, 135
plastics, 49
platinum, 74
pneumonia, 51, 52
pneumonitis, 52
pollen, 63
pollutants, 5, 8, 45, 47, 48, 52, 53, 54, 58, 59, 145
pollution, 6, 8, 12, 13, 42, 46, 47, 49, 55, 56, 59, 64, 95, 96, 133, 134
polycarbonate, 102
population, 21, 58
porosity, 89
power generation, 91
practical case study, 131, 142, 144
Predicted Percentage Dissatisfied (PPD), 4
pregnancy, 41
preservation, 133, 137, 143, 144, 145
prevention, 10, 14, 76, 99
principles, 24, 40, 62
productivity rates, 147
programming, 85, 86
project, 42, 145
proliferation, 10, 57
protection, 94, 134
proteolytic enzyme, 10
prototypes, 91
publishing, 96
pumps, 114, 131, 132
PVC, 8

Q

questionnaire, 63

R

radiation, 18, 19, 34, 35, 38, 39, 68, 70, 72, 74, 90, 94

Radiation, 33
radium, 50
radon, 50
reactions, 46, 133
real time, 133
reasoning, 39
recognition, 133
recommendations, iv, 4, 10, 134, 148, 155
reconciliation, 92
recovery, 60, 91
recycling, 10, 60
regression, 28, 120
regulations, 37
reinforcement, 49
reliability, 93
REM, 94
renewable energy, 95
requirements, 13, 63, 64, 82, 115, 156
researchers, xiii, 9, 83, 85, 115
resistance, 26, 40, 133, 145
resolution, 86
resources, 65, 81, 90
respiration, 24, 25
response, 4, 9, 15, 38, 54, 91, 115, 119, 132
risk, ix, xi, xiii, 10, 11, 33, 38, 45, 49, 51, 63, 76, 99, 144, 155
rodents, 50
room temperature, 134
rubber, 49
rules, 91, 94

S

safety, 147, 148
saturation, 22
savings, 8, 15, 21, 94
school, ix, xi, xiii, 29, 45, 108, 113, 115, 116, 117, 118, 120, 121, 122, 123, 124, 126
scope, 92
seasonal flu, 142
sensation, 4, 16, 18, 24, 27, 28, 29, 33, 35, 37, 55, 57, 151, 152

sensations, 16, 29
senses, 54, 72
sensing, 74
sensitivity, 32
sensors, 16, 32, 78
sex, 30, 151
shape, 7, 132
shock, 149
shortness of breath, 52
showing, 116
sick building syndrome (SBS), 46
signals, 16, 41
simulation, 12, 81, 82, 83, 84, 86, 89, 91, 92, 93, 94, 95, 96, 115, 116, 125, 129
simulations, 81, 90, 113, 115, 116, 119, 124, 126
skin, 3, 18, 21, 24, 25, 26, 32, 38, 42, 45, 50, 91, 128, 146, 151
smallpox, 51
smoking, 48
software, xiii, 65, 81, 82, 90, 91, 92, 95, 96, 113, 114
solution, 134, 144, 153
solvents, 49
sorption, 89
Spain, ix, xi, xiii, 101, 111, 113, 114, 116, 142
species, 133
specific heat, 25, 86, 88, 89
specifications, 152
spore, 11
spreadsheets, 21
stability, 145
standard deviation, 152
standardization, 9
state, 24, 85, 89, 94, 102, 104, 116, 118
statistics, 29
steel, 6
storage, 25, 82, 83, 92, 96, 116, 133, 134, 145
stoves, 48
stress, 40, 41, 73, 75, 76, 79, 80, 147, 148, 153, 156

structural modifications, 32
structure, 82, 85, 87, 89, 118, 125
substrates, 52
surface area, 25, 69, 89, 153
survival, 11
sweat, 3, 26, 38, 39, 73, 153
Sweden, 52, 97, 129
symptoms, 9, 42, 45, 51, 52, 104
syndrome, 46, 63, 64
system analysis, 89, 94

T

techniques, xiii, 11, 46, 59, 65, 85
technology, 95, 133, 145
textiles, 49, 133
thermal analysis, 91, 94, 95
thermal energy, 37, 84
thermal properties, 89
thermal resistance, 26, 29, 88
thermal stability, 116
thermoregulation, 3
time periods, 133
tissue, 50
tobacco, 13, 64
tobacco smoke, 13
total energy, 81
toxic substances, 45
training, 6
transactions, 42
transducer, 34, 68, 69, 70, 71, 72, 74, 75, 76, 77, 102
TRaNsient SYstems Simulation (TRNSYS), 84
transistor, 70
transmission, 51, 124
transparency, 85
transport, 91, 95
treatment, xiii, 91, 95
tuberculosis, 51
turbulence, 32, 33

U

UK, 28
unconditioned, 132
uniform, 18, 27, 34, 90, 108, 110
United, 52, 128
United States (USA), 52, 53, 92, 128, 156
University of A Coruña, ix, xi, 12, 157
urban, 11, 131, 132
urea, 49

V

validation, 6, 9, 92
variables, 4, 6, 21, 58, 150
variations, 34, 78, 135, 142
vehicles, 32, 80
velocity, 4, 13, 17, 18, 21, 27, 32, 33, 58, 71, 72, 74, 75
ventilation, xi, 10, 11, 15, 45, 46, 51, 53, 54, 55, 56, 57, 58, 59, 60, 62, 63, 64, 67, 75, 79, 86, 91, 93, 95, 96, 101, 102, 103, 104, 108, 110, 111, 114, 116, 118, 119, 125, 127, 132, 142, 143, 146, 153, 155
vessels, xiii
volatile organic compounds, 62

Volatile Organic Compounds (VOC), 78
vote, 16, 17, 25, 42, 147

W

Washington, 128
water, 10, 11, 22, 23, 26, 32, 40, 46, 47, 49, 51, 52, 75, 87, 89, 91, 135, 137, 155
water vapor, 47
web, 96
wetting, 133
windows, 31, 34, 60, 78, 89, 90, 94, 108, 131, 132, 135, 136, 139
Wisconsin, 84, 97
wood, 49, 126, 133
work environment, 59, 148
workers, 26, 38, 148, 149, 155
working conditions, 114
workload, 41
workplace, 41
World Health Organization (WHO), 46, 63
worldwide, xiii

Y

yield, 94, 153